Equilibrium Statistical Mechanics

E. Atlee Jackson

Dover Publications, Inc.
Mineola, New York

Published in Canada by General Publishing Company, Ltd., 30 Lesmill Road, Don Mills, Toronto, Ontario.

Bibliographical Note

This Dover edition, first published in 2000, is an unabridged republication of the work published by Prentice-Hall, Inc., Englewood Cliffs, NJ, in 1968.

Library of Congress Cataloging-in-Publication Data

Jackson, E. Atlee (Edwin Atlee), 1931–
Equilibrium statistical mechanics / E. Atlee Jackson.
p. cm.
Originally published: Englewood Cliffs, N.J. : Prentice-Hall, 1968.
Includes index.
ISBN 0-486-41185-0 (pbk.)
1. Statistical mechanics. I. Title.

QC175 .J23 2000
530.13'2—dc21

99-087129

Manufactured in the United States of America
Dover Publications, Inc., 31 East 2nd Street, Mineola, N.Y. 11501

To my wonderful triumvirate

Cynthia

Eric

Mark

Preface

This book is a result of teaching equilibrium statistical mechanics to upper-level undergraduate students. Before delving into statistical mechanics proper, the first two chapters are intended to introduce the student to statistical concepts and to review and extend his often rather vague appreciation of energy. The first chapter (probability) attempts to familiarize the student with such concepts as probability, statistical independence, distribution functions, averages, and statistical uncertainty. This material is usually new to students and should be covered thoroughly. The second chapter (energy) may, on the other hand, be used in a variety of ways, depending on the students' background. The early sections of this chapter are quite elementary, and may even be left as a reading assignment for the students (and many should read them!). The subsequent sections, concerning models of physical systems, are intended not only to introduce these models but also to get the student into a "microscopic" frame of mind — to think about atoms, the various forms of energy, and their orders of magnitude. Since most of these models are not actually used until Chapter 4, they may be initially treated in a cursory manner and referred to in more detail when the applications are studied. The purpose of discussing a number of models at this point is in the hope that it will help the student acquire more of an all-over picture of the variety and magnitude of forces, atomic arrangements, interactions, etc., which occur in nature. It also serves as an easily located (albeit, limited) "handbook" for all later applications. Finally, the last section of Chapter 2 simply contains a record of the more relevant results from quantum mechanics and a comparison of the orders of magnitude of various types of quantum "jumps." No derivation of these results is attempted in this book. If the student is interested in the origin of these results, references are supplied at the end of the chapter. I believe that this procedure is more realistic than attempting to give a one-page "derivation" of each result.

In Chapter 3 an attempt has been made to give a careful, simple, and yet general discussion of statistical mechanics. The method used is not very common, but it has the advantages of simplicity and generality and, moreover, circumvents all combinatorial analysis, Lagrange multipliers, most probable distributions, and the like. The fact that the fundamentals of statistical mechanics can be developed without pages of $N!/n!\,(N - n)!$ symbols is, in my opinion, a blessing which should be perpetuated. In the present approach these combinatorics are used, but as auxiliary tools rather than as essential elements in the foundations of statistical mechanics. The material in Section 8 of this chapter (open systems) is not used until Sections 14 to 16 of Chapter 4 and may be omitted or deferred until these applications are discussed.

The applications in Chapter 4 have been arranged (more or less) according to the amount of additional knowledge which must be brought to bear on the subject. Thus the first five sections involve only classical statistical mechanics. Sections 6 and 7 indicate one of the breakdowns of this classical theory, thereby motivating the introduction of quantum statistics. Sections 8 through 13 deal with applications of nondegenerate quantum statistics. Finally, Sections 14 through 16 treat the degenerate perfect gases.

The problems should be considered an integral part of the text. They vary from elementary applications (to give the student some confidence) to genuine extensions of the text which require some imagination. Their ordering is usually on this basis.

I am indebted to many of my hapless students for their suggestions concerning the presentation of this material. Any further comments or corrections would also be welcomed.

E. Atlee Jackson

Contents

Equilibrium Statistical Mechanics

Introduction

To understand the objectives of statistical mechanics, it is useful, by way of contrast, to review briefly the purpose and some of the results of classical thermodynamics. Classical thermodynamics is (from one point of view†) essentially a science concerned with establishing the relationships between the macroscopic properties of a system. For this purpose one must assume the existence of an equation of state that relates the state variables of a system. Then, using the very general laws of thermodynamics, one may establish other basic properties of the system (internal energy, entropy). Determining the relationships between these properties or between other derived properties (compressibility, expansiveness, heat capacities, Gibbs and Helmholtz functions, and so on) is then a relatively simple matter of applying the mathematical relationships between partial derivatives. The crowning concept of classical thermodynamics is entropy. Without this property it would be impossible to develop many of the useful relationships between the directly observable properties of the system. The usefulness of thermodynamics lies, in part, in its broad generality, which is a result of the fact that very few restrictive assumptions are made about the

†One hundred thermodynamicists will have 100 points of view!

systems under consideration. Of course "you get what you pay for" — and consequently, if only a few assumptions are made about a system, one cannot expect to predict any details about the properties of the system. Thus classical thermodynamics does not attempt to predict even the dependency of the macroscopic properties on the state variables (much less their numerical values). It is only capable of establishing the relationships *between* the properties of a system.

Statistical mechanics is a science that attempts to go beyond these results. For this purpose, however, one must make further assumptions about the system under consideration. First of all it is necessary to make assumptions about the *microscopic* composition of the system (such as the types of atoms, intermolecular forces, the magnitude of the dipole moments and magnetic moments, and so on). This extension to microscopic considerations is an important difference between statistical mechanics and classical thermodynamics. As soon as this step is made, one immediately faces the problem of dealing with a large number of particles (of the order of 10^{19} per cm^3). Experience shows that the *macroscopic* properties of a system do *not* depend on what *each* atom is doing (e.g., where *each* atom is located and how fast *each* atom is moving). If there did exist such a dependency, then there would be no science of classical thermodynamics, for thermodynamics does not take these facts into account. Rather, the macroscopic properties of a system must depend only on some *average* behavior of all the atoms. This fact is extremely important. If it were not true, then we would have to predict what each atom (10^{19} of them!) is going to do at each instant in time. This would be impossible, even with the help of the best computer. Thus, we must take full advantage of the lesson taught by experience, namely:

The physical properties of a macroscopic system depend only on the average behavior of all the atoms in that system.

We are still left with the problem of how to predict the average behavior of many atoms — and here we must consider statistical methods. Statistical methods are particularly adapted for treating *just* those situations in which it is impossible to predict events with certainty. Since we do not know the positions and velocities of all the atoms, we clearly cannot predict the future configuration of atoms with certainty (nor do we need to!). But we *can* make reasonable statistical assumptions concerning macroscopic systems. This, then, is the program of statistical mechanics:

1. Make a reasonable statistical assumption about *all* systems and, using the methods of probability, obtain *general* expressions for the macroscopic properties.

2. Use further assumptions about interatomic forces and the like for any particular system to evaluate the general expressions already obtained. In this way one can obtain the dependency of the properties of that system on the state variables and also obtain approximate numerical values for these properties.

Obviously, in order to accomplish this program, one must acquire some familiarity with probability and with molecular concepts. A vast knowledge is not required, but only an acquaintance with a few basic ideas. Thus we must first acquire the needed tools of the trade, and then apply them to physical systems to obtain the science of statistical mechanics. This we shall now proceed to do.

1

Probability

1. FREQUENCY AND PROBABILITY

It is a common experience that the outcome may not always be the same when an experiment is performed a number of times, even though the conditions of the experiment are kept as similar as possible. The reason is that some of the factors that contribute to the outcome of the experiment are not (or cannot) be completely controlled. Simple examples are the "experiments" of rolling dice, drawing cards, tossing coins, or any of the so-called games of chance. Presumably other experiments are nearer to the hearts of physical scientists, but these examples will suffice for the present. In any case, the typical feature of all experiments is that at the end of the experiment one observes some result of interest. To be concise, we shall call those distinct (or mutually exclusive) results of an experiment that are of interest **simple events.** Therefore the result of each experiment is always one, *and only one,* simple event. For simplicity we may label these simple events (or simply "events") with some index i. Thus the two possible events when tossing a coin are heads or tails ($i = h, t$), whereas there are six possible events when a single die is rolled ($i = 1, 2, 3, \ldots, 6$), and so on.

Now if a particular experiment is performed a number of times, say N times, a particular event i may be found to occur n_i times. This fact is of considerable interest, because if the experiment is repeated at a later time, we expect the event i to occur with roughly the same frequency. To investigate this idea we consider the ratio

$$F_i = \frac{n_i}{N} \tag{1}$$

This ratio is the fraction of the N experiments that resulted in the event i and is commonly called the *frequency* of the event i. Although it is useful to know the value of F_i found in some previous group of N experiments, it is important to realize that, if these N experiments are repeated, one cannot expect that the event i will occur the same number of times (n_i). Instead it may occur m_i times. This means that F_i will in general be different for different groups of experiments. Thus, for example, if a coin is tossed twenty times ($N = 20$), the event "heads" may occur eight times ($n_h = 8$), so that $F_h = 0.4$ *for that sequence* of tosses. If we tossed the coin again twenty times, we would consider it unlikely that heads would turn up again eight times, so we would expect a different value for F_h. Moreover, if the coin were tossed 100 times, the coin might turn up heads 54 times, in which case $F_h = 0.54$ for that sequence of tosses. If $N = 1{,}000$, we might observe $n_h = 510$, in which case $F_h = 0.51$. Clearly the frequency of an event depends on the group of experiments being considered.

Since the frequency of an event varies from one group of experiments to another, it is desirable to obtain a quantity that does not depend on any particular group and that at the same time indicates the frequency we can expect in any particular group of experiments. To obtain such a quantity we could, at least in principle, examine the values of the frequency as N becomes extremely large. In the above examples we had

$$F_h(N = 20) = 0.4, \qquad F_h(N = 100) = 0.54, \qquad F_h(N = 1{,}000) = 0.51$$

As N becomes larger and larger we expect that, if the coin is evenly balanced, the frequency F_h will approach the value 0.50. However, regardless of what the limiting value of the frequency may be when N becomes extremely large, we call this limiting value the *probability* of a heads (for that coin). Thus, in principle, we have a method for obtaining a quantity P_h (the probability of a heads) that is related to the frequency we can expect to find in future experiments.

We now can give a formal definition for the **probability** P_i of an event i, namely,

$$P_i = \lim_{N \to \infty} F_i = \lim_{N \to \infty} \frac{n_i(N)}{N} \tag{2}$$

By this we mean that P_i equals the limiting value of F_i as N becomes arbitrarily large. This in turn equals the limiting value of n_i/N, where n_i also depends on the value of N.

There are two ways of interpreting Equation (2), both of which are frequently used in statistical mechanics. We have said that N represents the number of experiments, and n_i is the number of these experiments that result in the event i. Now this can be interpreted in two ways. First, we can picture *one* physical system on which we perform the same experiment over and over again (altogether, N times). The number n_i is, in this case, the number of times the event i occurs in this sequence of experiments. From this point of view the experiments are carried out at different times, one after the other. In practice, this is usually what one does. However, there is another interpretation of Equation (2) into which time does not enter. In the second method we envisage N identical systems (for example, N identical coins, N decks of cards, or N bottles of a gas). The systems are identical in the sense that we cannot distinguish between them by any *macroscopic* method. Such a collection of identical systems is called an **ensemble.** Now we perform the same experiment on *each* of the N systems of this ensemble and take n_i to be the number of these systems that yield the event i. For example, we might have an ensemble of 1,000 identical coins. All of these coins are now flipped (at the same time, if you like), and it is found that 510 turn up heads. Then $n_h = 510$, so the frequency is $F_h = 0.51$. From this point of view the time when the experiments are performed is clearly unimportant in determining F_h, whereas in the first interpretation it is not so clear how time might affect the answer. *We shall assume from now on that either method will yield the same result.* (In statistical mechanics this assumption is known as the ergodic hypothesis.) Thus we can think of frequency (and probability) in terms of either a sequence of experiments on one system or one experiment on each member of an ensemble. Time, therefore, plays no role in determining the probability.

Although definition (2) is fine in principle, in practice one repeats an experiment only a finite number of times. Alternatively, one can only construct an ensemble containing a finite number of systems. Thus, in either case, the limit $N \to \infty$ in Equation (2) cannot be realized in any physical situation. For this reason one can obtain only an approximate value for the probability of an event. Nonetheless, one is likely to say that the probability of tossing a heads is $\frac{1}{2}$, or the probability of picking a particular card from a deck is $\frac{1}{52}$. A statement of this sort is based on the *assumption* that the probabilities of certain events are equal. To see how such numbers (i.e., $\frac{1}{2}$, $\frac{1}{52}$) result from such assumptions, we must establish two properties of the probability (and the frequency). First, the frequency of any event is clearly a positive number, and consequently, because of definition (2), all P_i are positive numbers. Second, if a coin is tossed N

times and the coin turns up heads n_h times, it clearly must turn up tails $n_t = N - n_h$ times. Put another way, we must have $n_h + n_t = N$, because in each experiment one of the events, heads or tails, must have occurred. For the same reason we must find that the sum of n_i for all possible events i equals the total number of experiments N. This means that the sum of all F_i must equal unity (no matter what the value of N may be), and consequently the sum of all P_i must also equal unity. We therefore have the two simple but very important properties of the probability:

$$\begin{aligned} P_i &\geq 0 \qquad \text{(for all simple events } i\text{)} \\ \sum_i P_i &= 1 \qquad \text{(summed over all simple events)} \end{aligned} \tag{3}$$

Now we shall see how these properties (actually only the second), together with certain assumptions, yield values for the probability of certain events. In the case of a coin we expect that the probability of a heads P_h and that of a tails P_t are equal. If we make this *assumption*, then we can set P_h and P_t equal to their common probability (call it P_0), in which case

$$P_h + P_t = 2P_0 = 1 \qquad \text{or} \qquad P_0 = \tfrac{1}{2} = P_h = P_t$$

This shows how the value of $\frac{1}{2}$, quoted above, is arrived at once one has made an assumption about the equality of the probabilities of certain events. It should be emphasized that if we have no information about the coin, the assumption that P_h equals P_t is the most reasonable assumption to make. Although the present example is nearly trivial, it contains all (or nearly all) of the essential features used in more complicated cases.

To illustrate these points with another somewhat more complicated example, consider the experiment of drawing a card from a deck of cards. In this case there are 52 possible events, each with some probability $P_i (i = 1, 2, 3, \ldots, 52)$. If we have no information about the fact that the cards are marked or that there is some legerdemain being used, *the most reasonable assumption to make* is that the probability of drawing any particular card is equal to the probability of drawing any other card. Let us call this common probability P_0 (again). Then, using (3), we have

$$P_1 + P_2 + P_3 + \cdots + P_{52} = \sum_i P_i = 52P_0 = 1 \qquad \text{or} \qquad P_i = P_0 = \tfrac{1}{52}$$

in agreement with our statement (above) that the probability of drawing a particular card from a deck is usually assumed to be $\frac{1}{52}$. What is important here is to realize what assumptions are being made and how these assumptions yield a value for the probability.

The procedure outlined above is frequently used. That is, to *predict* the probability of a certain event, one uses the general rule:

If there is no apparent reason for one event to occur more frequently than another, then their respective probabilities are assumed to be equal. (4)

Probabilities obtained from the reasoning in (4) should be more accurately termed "a priori probabilities" (before, and hence independent of, experience), whereas those obtained by $P_i = \lim_{N \to \infty} n_i(N)/N$ could be called "a posteriori probabilities" (or empirical probabilities); however, the terminology is cumbersome and will not be used. Notice, nonetheless, that a physical theory involving probability is usually based on "a priori probability." The final justification of the theory rests on the agreement between the predicted results of the theory and the observed experimental results.

In this section we have discussed how probability is usually related to the observed frequencies in physical experiments. Moreover, we have seen how general arguments [Equation (4)] are frequently used to *predict* the probability of various events. We shall now consider some of the more formal aspects of probability, which allow us to determine the probability of more complicated events.

2. *PROBABILITY OF COMPOUND EVENTS: INDEPENDENT EVENTS*

It will be recalled that the result of an experiment is always one, and only one, simple event. The probability of these events was denoted by P_i, and they have the properties

$$P_i \geq 0$$
$$\sum_i P_i = 1 \qquad \text{(summed over all simple events)} \tag{5}$$

The assignment of numerical values to the various P_i is, in practice, usually based on an argument such as (4). However, in principle, the values may be assigned for any reason, provided only that they satisfy (5).

Now it is very helpful to think of these events as points in a space, called the "sample space." Such a space is illustrated in Figure 1. It is simply a collection of points, each of which represents a possible result of an experiment. In this space we are not interested in distances, or the arrangement of the points, but only in the points themselves.

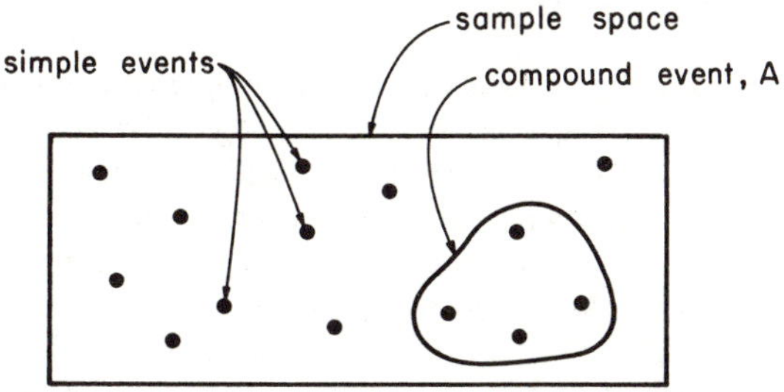

Figure 1

By a *compound event* we shall mean any specified collection of the points in the sample space. The collection of points in the compound event are specified by some feature they have in common (as we shall illustrate shortly). In Figure 1 a compound event A has been indicated. It includes all the points enclosed by the boundary. The probability of this compound event $P(A)$ is *defined* to be

$$\boxed{P(A) = \sum_{i \subset A} P_i} \tag{6}$$

where the sum is over all the points in the event (indicated by the symbols "$i \subset A$," which reads "all the i's contained in the event A").

To illustrate these points, consider the experiment of drawing a single card from a deck of cards. The possible simple events are fifty-two in number, and can be labeled $i = 1, 2, \ldots, 52$. Our sample space in this case consists of fifty-two points, each representing one of the cards that may be drawn. Now we define three compound events A, B, and C that we might be interested in. These events are specified by certain features that all the points have in common. Thus:

A: All points that represent hearts

B: All points that represent a number three card

C: All points that represent a one-eyed jack

On the basis of (4) we shall assign equal probability to each of the points in the sample space (i.e., to each simple event). Because of (5) we conclude, as we did in the last section, that the probability of each simple event equals $\frac{1}{52}$. Now we determine the probability of each of the compound events above, using definition (6). We have

$$P(A) = \sum_{i \subset A} P_i = 13 \times (\tfrac{1}{52}) = \tfrac{1}{4}$$

for there are 13 points representing a heart, each with probability $\frac{1}{52}$. Similarly,

$$P(B) = \sum_{i \subset B} P_i = 4 \times \tfrac{1}{52} = \tfrac{1}{13}, \qquad P(C) = \tfrac{2}{52} = \tfrac{1}{26}$$

That is, the probability of drawing a heart is $\frac{1}{4}$, of drawing a number three card is $\frac{1}{13}$, and of drawing a one-eyed jack is $\frac{1}{26}$. This sample space and the compound events A, B, and C are illustrated in Figure 2.

Two compound events may or may not have points in common. Thus, in Figure 2, events A and B have a single point in common (namely, the point corresponding to the three of hearts), whereas events B and C have no points in common (there is no number three card that is also a one-eyed jack). If two events do not have any points in common, we say that they

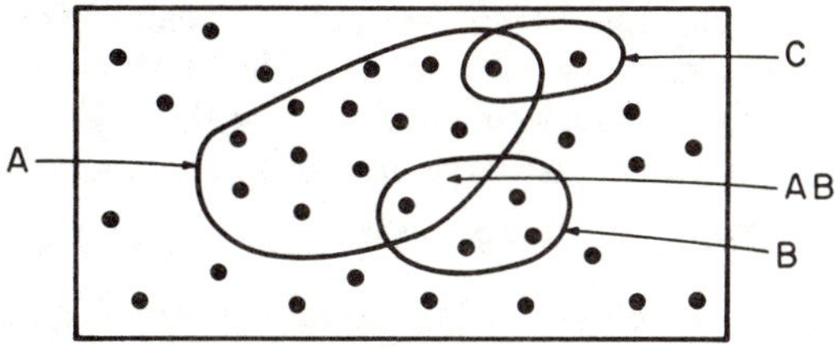

Figure 2

are *disjoint.* Now consider the disjoint events B and C. We may be interested in knowing the probability that the outcome of the experiment will be either event B or event C (i.e., what is the probability that the card drawn is either number three or a one-eyed jack?). This is, of course, just another compound event, which we can represent by $B \cup C$ (read: B union C). The *union* of B and C is simply the compound event consisting of all points that are either in B or in C (or in both B and C). The probability of this event $P(B \cup C)$ is, according to (6), just the sum of the probability of all points in that event. Since B and C have no points in common (disjoint), this equals the sum of $P(B)$ and $P(C)$, or

$$P(B \cup C) = P(B) + P(C) \qquad \text{(if } B \text{ and } C \text{ are disjoint)} \tag{7}$$

More generally, if A and B are two events that are not disjoint, then $P(A \cup B)$ is still equal to the sum of the probabilities of all points in A and B. However, in order to express $P(A \cup B)$ in terms of $P(A)$ and $P(B)$ we cannot use (7). The reason is that the sum of $P(A)$ and $P(B)$ adds the probabilities of the points that A and B have in common *twice.* To rectify this double counting, we clearly need to subtract something. What we need to subtract is just the probability of all the points that A and B have in common. We therefore define the *intersect* of A and B as that event which contains all points common to both A and B and write it as AB. The intersect and the union of two events A and B are illustrated in Figure 3.

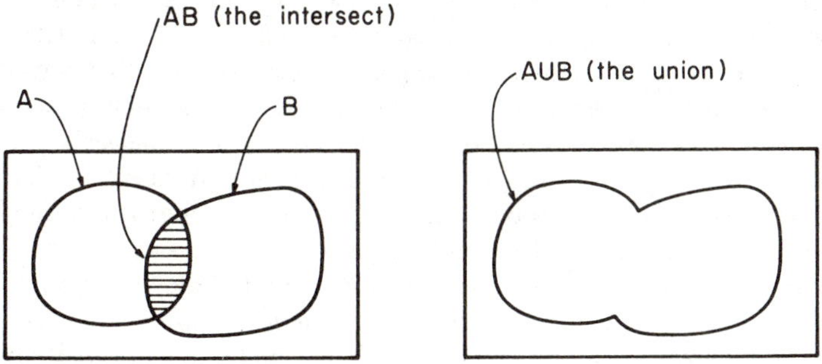

Figure 3

In the present example $P(AB) = \frac{1}{52}$, the probability of drawing a three of hearts. In the case of events B and C, which have no points in common, we have, according to Equation (6), $P(BC) = 0$, for there are no terms in the sum. Then, in general, we have for any two compound events X and Y

$$P(X \cup Y) = P(X) + P(Y) - P(XY), \tag{8}$$

which reduced to $P(X) + P(Y)$ if X and Y are disjoint, $P(XY) = 0$.

In our example

$$P(B \cup C) = \tfrac{1}{13} + \tfrac{1}{26} = \tfrac{3}{26}$$

for there are six cards in the deck that are either one-eyed jacks or numbered three. On the other hand

$$P(A \cup B) = \tfrac{1}{4} + \tfrac{1}{13} - \tfrac{1}{52} = \tfrac{4}{13}$$

for there are sixteen cards that are either hearts or numbered three.

The important new feature that occurs in the case of compound events (as contrasted with simple events) is that two compound events may have points in common (their intersect). Simple events, on the other hand, do not have points in common (by definition the result of an experiment is one, *and only one*, simple event). Thus simple events are always disjoint. To determine the probability of the union of some simple events (i.e., a compound event), we simply add up the relevant probabilities, using the definition $P(A) = \sum_{i \subset A} P_i$, Equation (6). We see from Equation (7) that there is really no essential difference between *disjoint* compound events and simple events, for we add up probabilities in the same way in both cases. It is only when compound events have an intersect that we must use some care. Then we found that (8) is the appropriate way to add probabilities. From now on we can refer to both compound and simple events as "events" and always use Equation (8). If the events happen to be disjoint, then it automatically reduces to (7).

We have introduced the concepts above so that we can define two final, very important notations that are used repeatedly in statistical mechanics. In fact, the whole purpose of this section is to arrive at the notion of **independent events.** To make the concept precise we must first consider the idea of **conditional probability.** Conditional probability measures the effect (if any) on the occurrence of an event A when it is *known* that an event B has occurred in the experiment. For example, we might ask: What is the probability that the card drawn is a heart if we *know* that it is a number 3? A minute's thought shows that its being a three has no effect on the probability that it is a heart. In other words, this probability should still be $\frac{1}{4}$. On the other hand, we might ask: What is the probability that the card is a heart if we *know* it is a one-eyed jack? Now the fact that it *is* a one-eyed jack limits the possible suits to either hearts or spades; consequently, the above probability would be $\frac{1}{2}$ (using our usual assumption). Clearly we now have an "effect" on the probability of drawing a heart when we know the event C occurred. We now *define* the **conditional probability** $P(X \mid Y)$ by

$$\boxed{P(X \mid Y) = \frac{P(XY)}{P(Y)}} \tag{9}$$

By $P(X \mid Y)$ we mean the probability that event X occurs given the fact that event Y occurred in the experiment. Clearly if events X and Y have no points in common (disjoint) so, that $P(XY) = 0$, event X cannot occur if event Y occurred, which agrees with definition (9). Moreover (9) agrees with our two examples above, for

$$P(A \mid B) = \frac{\frac{1}{52}}{\frac{1}{13}} = \frac{1}{4}$$

$$P(A \mid C) = \frac{\frac{1}{52}}{\frac{1}{26}} = \frac{1}{2}$$

Thus, though (9) is a definition that is good for all cases, it is nice to see that it agrees with our intuition when our intuition is sufficiently clear to give an answer! We note that in general $P(X \mid Y)$ is not equal to $P(Y \mid X)$. Thus $P(B \mid A) = \frac{1}{13}$, for our above example.

We note that in our example with cards

$$P(A \mid B) = P(A) \qquad \text{and} \qquad P(B \mid A) = P(B) \tag{10}$$

which, according to (9), is another way of saying that

$$\boxed{P(AB) = P(A)P(B)} \tag{11}$$

When such a relationship holds between two events A and B, we say that the two events are **independent.** As we have already seen, not all events are independent. In the example above

$$P(AC) = \tfrac{1}{52} \neq P(A)P(C) = (\tfrac{1}{4})(\tfrac{1}{26}) = \tfrac{1}{104}$$

Calling events A and B independent makes sense from the point of view that if one occurs during an experiment, it does not affect the probability that the other will occur.

The essential point to understand (because it will be crucial in the development of statistical mechanics) is this concept of independence. By definition, two events A and B are independent only if Equation (11) is satisfied, and this in turn is satisfied only if Equation (10) is valid. This then gives a precise definition to independence, but it is helpful to state its meaning in words (even if the meaning is not precise).

If the occurrence of one event does not affect the chances of the occurrence of another event, then these two events are said to be independent. (12)

3. *DISTRIBUTION FUNCTIONS*

The principal use of a sample space is to help us order our thoughts when analyzing problems. Like all devices it has its limitations, and there are many cases in which it is not very useful. In particular, if the points in a sample space have different probabilities, then we need a more convenient method of recording these data than to write the probabilities beside each point. As an example, consider rolling two dice. In some social circles the event of interest is the sum s of the numbers on the two upturned faces. Assuming that each side of both dice is equally likely to turn face up at the end of a roll, one can show (can you?) that the probability $P(s)$ of the sum s is:

s	2	3	4	5	6	7	8	9	10	11	12
$P(s)$	$\frac{1}{36}$	$\frac{1}{18}$	$\frac{1}{12}$	$\frac{1}{9}$	$\frac{5}{36}$	$\frac{1}{6}$	$\frac{5}{36}$	$\frac{1}{9}$	$\frac{1}{12}$	$\frac{1}{18}$	$\frac{1}{36}$

In this case the sample space has eleven points, with six different values of the probability. Although the sample space is still useful for analyzing problems, it is not appropriate for recording these values of $P(s)$. Of course, the table above is one way of recording them. There is another way, however, which has the visual advantages of pictorial representation: namely, a graph. Such a graph is shown in Figure 4.

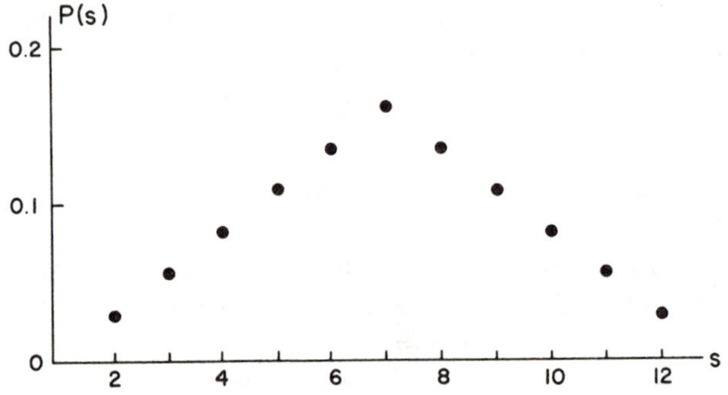

Figure 4

When $P(s)$ is represented in this manner, it emphasizes that the probability $P(s)$ is actually a *function* of s. This is clearly true in general. That is, the probability $P(i)$ of an event that is characterized by the number i is actually a *function* of i, and is called a *probability distribution function*. The name comes from the fact that $P(i)$ determines how the probability is distributed over the events i. This function, of course, has the properties

$$P(i) \geq 0, \qquad \sum_i P(i) = 1 \tag{13}$$

One should not look for anything particularly deep here — $P(i)$ is still the probability of the event i. We are simply taking note of the fact that $P(i)$ can be considered a function of i, which tells us how the probability is "distributed." As usual, functions can most easily be represented by a graph of the type shown in Figure 4.

We shall now consider an important generalization of the probability distribution function discussed above. Up to now, for simplicity, we have discussed fairly simple "experiments." A more realistic experiment might involve a gas in a container that has a fine slit in it. The gas atoms escape from the container into a vacuum, which is maintained by a vacuum pump. We would like to know the probability that a gas atom (which escapes from this container) is traveling in a certain direction. To determine this, we could set up a fluorescent screen as shown in Figure 5. Every time an atom

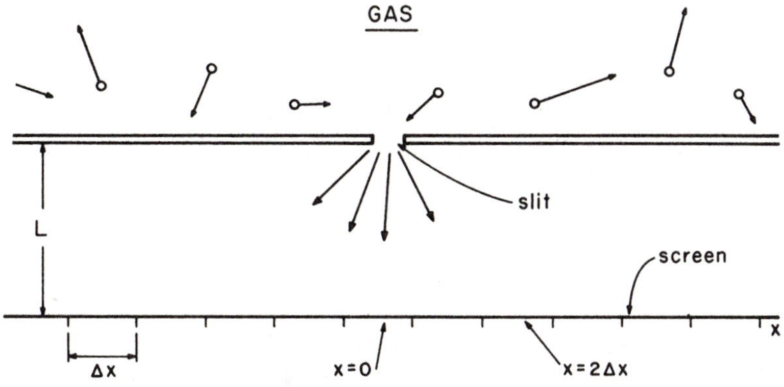

Figure 5

hits the screen it causes the screen to scintillate, and thus we can determine the direction in which that atom was traveling (for we know it came from the slit). The first thing that becomes apparent is that there is, in this case, a *continuous range* of possible events — namely, an atom can strike any *point* on the screen. It is clear that the chances are nil for any atom to strike a particular (mathematical) point, regardless of how long we watch this system. This shows that our original question has a very simple (but not very useful) answer. The "chance" of finding an atom traveling in a particular direction is simply zero! What is it that we really want to know? It is really not important whether or not an atom is moving in some *precise* direction. All that we are interested in is the chance of finding it to be moving in "roughly" a certain direction.

To determine this we need only divide the screen up into strips of finite width Δx (e.g., 1 cm) running parallel to the slit. Whenever an atom strikes *anywhere* in that strip, we shall count it. The possible results of our experiment can now be labeled by the strip that was struck by the atom (the strips now constitute the events in our sample space). To experimentally determine the probability (or frequency) that an atom will strike a certain strip, we must record the number of hits $N(x)$ in the strip whose center is at x (see Figure 5) and divide by the total number of atoms that struck the screen (recall the definition of frequency in Section 1). This record of the number of hits can be conveniently recorded by a histogram (see Figure 6).

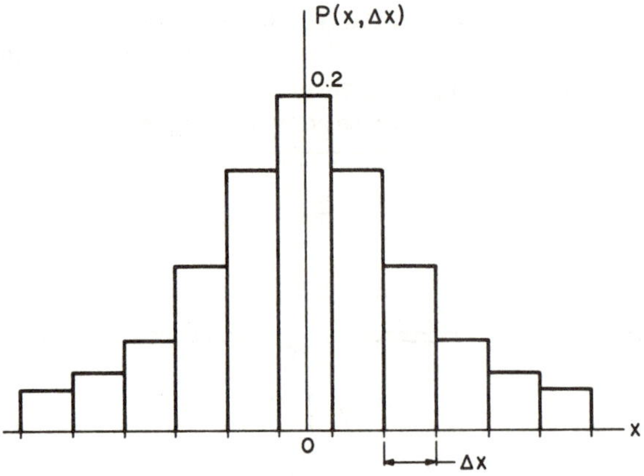

Figure 6

The height of each bar represents the number of hits in that strip, divided by the total number of atoms observed N. If N is very large, then this becomes the probability.

$$P(x, \Delta x) = \begin{array}{l}\text{probability that an atom will strike the strip of} \\ \text{width } \Delta x \text{, whose center is at } x\end{array} \tag{14}$$

If a person were more ambitious and wanted to determine the probable direction of the atoms more accurately, he might use smaller-sized strips. If the strips were half as big as before, he might obtain a histogram as shown in Figure 7. The information thus recorded is more detailed than in the previous case. Note that the number of atoms striking each strip is smaller, and hence *the probability of an atom's striking any particular strip is also smaller.* This agrees with our previous observation that the probability that an atom will move in a particular direction is zero (corresponding to the case in which the strip width is reduced to zero, or simply a line).

Now as the strip width is reduced, it is clear that this histogram becomes more continuous and can be approximated by a continuous function of x. However, since $P(x, \Delta x)$ goes to zero as Δx goes to zero, we consider the function

$$f(x) = \lim_{\Delta x \to 0} \frac{P(x, \Delta x)}{\Delta x}$$

or simply

$$\boxed{f(x)\,dx = P(x, dx)} \tag{15}$$

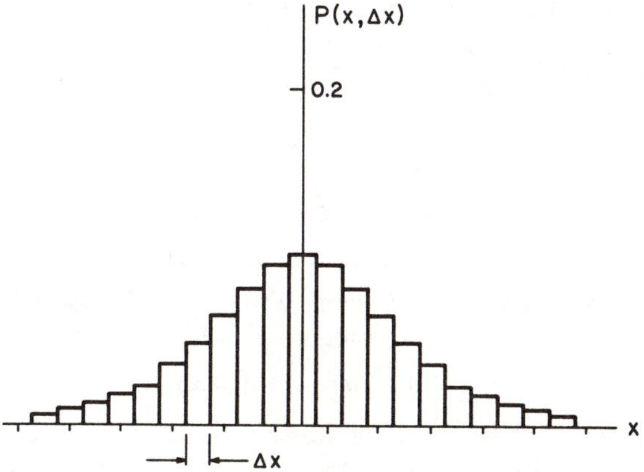

Figure 7

where *dx is understood to be an infinitesimal.* The function $f(x)$ is called a *probability density distribution function,* or simply a **distribution function.** The latter name is the one most commonly used in statistical mechanics and will be used from now on. However, it should be emphasized that $f(x)$ *is not the probability that an atom will strike at* x (for this is zero — corresponding to $dx = 0$). Instead, $f(x)$ represents the distribution of *probability densities,* so that $f(x)\,dx$ is the quantity that yields the probability. [See Equations (14) and (15).] These points are illustrated in Figure 8.

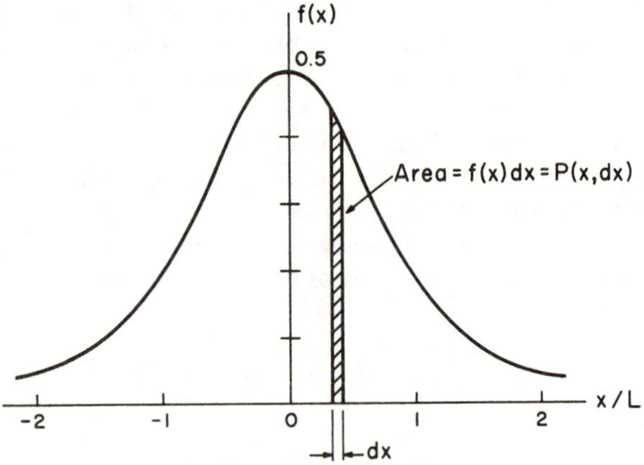

Figure 8

Using the fact that the sum of the probabilities must equal unity and the definition $f(x)\,dx = P(x, dx)$, we have

$$\sum_{x=-\infty}^{\infty} P(x, dx) = \sum_{x=-\infty}^{\infty} f(x)\,dx = \int_{-\infty}^{\infty} f(x)\,dx$$

so

$$\boxed{\int_{-\infty}^{\infty} f(x)\,dx = 1} \tag{16}$$

A function that satisfies (16) is sometimes said to be *normalized to 1*.

If N is the total number of atoms that strike the screen, the function

$$F(x)\,dx = Nf(x)\,dx \tag{17}$$

represents the probable *number* of atoms that strike the screen in the region dx of the point x. In this case $F(x)$ is a function that is normalized to N, for

$$\int_{-\infty}^{\infty} F(x)\,dx = N\int_{-\infty}^{\infty} f(x)\,dx = N \tag{18}$$

Such functions are often used in statistical mechanics and are also referred to as distribution functions. Whereas $f(x)$ represents the distribution of probability densities, $F(x)$ represents the distribution of the *probable number* densities. The essential difference between these distribution functions is only their normalized values (16) and (18).

The distribution function $f(x)$ can be used to compute the probability of compound events in the same way as the probabilities $P(x, dx)$. Thus we might want to determine the probability that an atom will strike between two points x_1 and x_2. This just represents a *compound event* — namely, the collection of all those events (x, dx) for which $x_1 \leq x \leq x_2$. The probability of this compound event is [according to the definition $P(A) = \sum_{i \subset A} P_i$]

$$\sum_{x_2 \geq x \geq x_1} P(x, dx) = \int_{x_1}^{x_2} f(x)\, dx$$

Of course, if x_2 is very close to x_1, then $f(x)$ is essentially constant over the range of integration, so

$$\int_{x_1}^{x_2} f(x)\, dx \simeq f(x)(x_2 - x_1) = f(x)\,\Delta x \qquad (x_2 \geq x \geq x_1)$$

and we return to the result $P(x, dx) = f(x)\, dx$ (when $\Delta x \to dx$).

If the slit were replaced by a pinhole in this experiment, then it would be more appropriate to divide the screen into rectangular cells of area $dx\, dy$ rather than strips. By using cells, we can now gain more information about the direction in which the atoms are moving, for we have a point source. The probability that an atom would strike such a cell would then be represented by

$$\boxed{P(x, y, dx, dy) = f(x, y)\, dx\, dy} \tag{19}$$

where (x, y) is the center of the cell. The cells on the screen are illustrated in Figure 9.

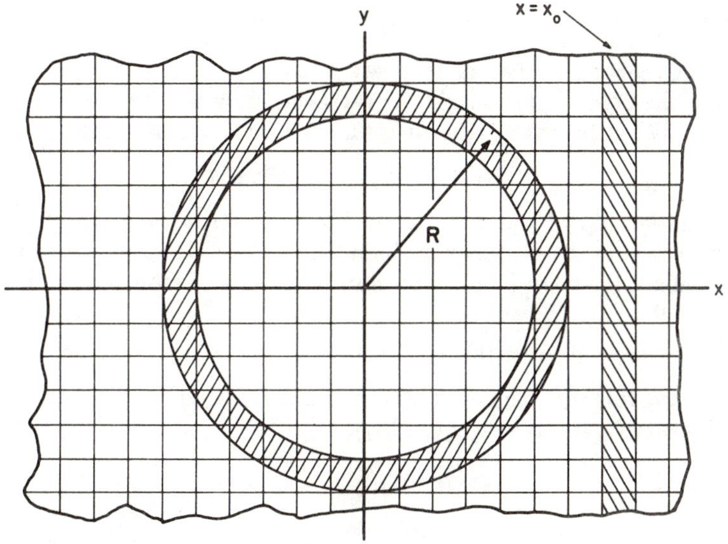

Figure 9

The distribution function $f(x, y)$ is now a function of the two variables x and y. From this distribution we can, as before, determine the probability of other (compound) events. For example, if

$$f(x, y) = \frac{L^2}{(L^2 + x^2 + y^2)^2} \frac{1}{\pi} \tag{20}$$

then the probability that an atom would hit a strip dx wide at the point $x = x_0$ would be given by adding up all the events with this value of x, namely,

$$f(x_0)\, dx = dx \int_{-\infty}^{\infty} dy\, f(x_0, y) \tag{21}$$

This is illustrated by the shaded strip in Figure 9. If $f(x, y)$ is given by (20), then $\left(\text{note that} \int \frac{dy}{c^2 + y^2} = \frac{1}{c} \tan^{-1} \frac{y}{c}\right)$

$$f(x_0)\, dx = dx \frac{L^2}{\pi} \int_{-\infty}^{\infty} \frac{dy}{(L^2 + x_0^2 + y^2)^2} = \frac{L^2}{2(L^2 + x_0^2)^{3/2}}\, dx$$

Clearly a similar expression could be obtained for $f(y)\, dy$ — the probability that an atom hits a strip of width dy about y.

Instead of the rectilinear coordinates (x, y) it is often more convenient to use cylindrical coordinates (r, θ), particularly if the situation has cylindrical symmetry. Here r is the distance from the central point on the screen and θ is the angle from the x axis. Thus

$$x = r \cos\theta, \qquad y = r \sin\theta$$

Now the probability that an atom will strike the screen in a region $(dr, d\theta)$ around the point (r, θ) can also be represented by a distribution function $F(r, \theta)$. We *define* this distribution function by the relationship

$$P(r, \theta, dr, d\theta) = F(r, \theta)\, dr\, d\theta \tag{22}$$

The relationship of the distribution functions $f(x, y)$ and $F(r, \theta)$ can be easily determined. Consider an infinitesimal area $dx\, dy = r\, dr\, d\theta$ on the screen about the point (x, y). The probability that an atom strikes this area is

$$f(x, y)\, dx\, dy \equiv f(r \cos\theta, r \sin\theta)\, r\, dr\, d\theta = P(r, \theta, dr, d\theta)$$

Comparing this with Equation (22), we see that $F(r, \theta) = rf(r \cos\theta, r \sin\theta)$. As an example, if $f(x, y)$ is given by (20), then

$$F(r, \theta) = \frac{rL^2}{\pi(L^2 + r^2)^2} \tag{23}$$

We see that $F(r, \theta)$ is independent of θ because the system has cylindrical symmetry. This makes integration over θ very simple. For example, the probability that an atom strikes in a ring of width dr about the radius $r = R$ (see Figure 9) is given by the sum of the probabilities of all events satisfying these conditions. Thus the probability is

$$dr \int_0^{2\pi} F(R, \theta) \, d\theta$$

and, if $F(r, \theta)$ is given by (23), then this simply equals

$$\frac{2RL^2}{(L^2 + R^2)^2}.$$

We have given examples of distribution functions containing one or two variables, but in many situations it is necessary to use more than two variables. Thus, for example, we might ask what the probability is that an atom has a velocity with an x component in the range dv_x of some value v_x, *and* a y component in the range dv_y of v_y, *and* a z component in the range dv_z of v_z. The required probability can be expressed in terms of a distribution function with three variables (v_x, v_y, v_z), such that

$$P(v_x, v_y, v_z, dv_x, dv_y, dv_z) = f(v_x, v_y, v_z) \, dv_x \, dv_y \, dv_z$$

The last expression can be written in the more compact notation $f(\mathbf{v}) \, d^3v$. An example of a distribution function of this type is the Gaussian distribution, $f(\mathbf{v}) = (\alpha/\pi)^{3/2} \exp(-\alpha v^2)$, where $\mathbf{v}^2 = v_x^2 + v_y^2 + v_z^2$. [Question: Why is there a factor $(\alpha/\pi)^{3/2}$ in this distribution function? See Appendix A and the examples in the following section.] We may also have distribution functions with more than three variables. For example, the probability that one atom has a velocity in the range d^3v_1 of $\mathbf{v}_1$, while (at the same time) another atom has a velocity in the range d^3v_2 of $\mathbf{v}_2$, could be represented by $f(\mathbf{v}_1, \mathbf{v}_2) \, d^3v_1 \, d^3v_2$. The distribution function would now be a function of six variables. An example of such a distribution function is $f(\mathbf{v}_1, \mathbf{v}_2) = (\alpha/\pi)^3 \exp[-\alpha(\mathbf{v}_1^2 + \mathbf{v}_2^2)]$. Clearly this process could be continued for more atoms.

This completes our introduction to distribution functions. We have seen a few examples of how we can use them to determine the probability of various events. Now we shall show how other quantities of interest are determined with the help of distribution functions.

4. *AVERAGES, VARIANCE*

Using the probability $P(i)$ of the events of an experiment, one can determine other quantities that are of interest. As a simple example, consider again the rolling of two dice. If these dice are rolled five times, one might obtain the sums 5, 9, 6, 10, and 6. The *average value* of these numbers is

$$\frac{5 + 9 + 6 + 10 + 6}{5} = \frac{36}{5} = 7.2$$

More generally, if the dice are rolled N times and the sum 2 occurs n_2 times, the sum 3 occurs n_3 times, and so on, then the average value for *those N rolls* is

$$\frac{1}{N}(2n_2 + 3n_3 + \cdots + 12n_{12}) = \sum_{k=2}^{12} k\frac{n_k}{N}$$

In the limit, as N goes to infinity, n_k/N becomes $P(k)$, so the right side becomes

$$\boxed{\sum_k kP(k) \equiv \bar{k}} \tag{24}$$

This quantity $\sum kP(k)$ gives the **average** (or mean, or expected) value of k, which we denote by $\bar{k}$. In the simpler case, when only one die is rolled, the average value is

$$\sum_{k=1}^{6} kP(k) = 1(\tfrac{1}{6}) + 2(\tfrac{1}{6}) + \cdots + 6(\tfrac{1}{6}) = 3.5$$

assuming that all sides are equally probable. The more complicated case of two dice is left as a problem.

We might not be interested in the average value of the sums that appear on the dice, but rather in some other quantity such as the square of the sum. By the same reasoning as above, the average value of the square of the sums would be given by

$$\overline{k^2} = \sum_k k^2P(k)$$

The quantity $(\overline{k^2})^{1/2}$ is called the *root mean square* of k and is sometimes written k_{rms}. In general, we define the average of *any* function of k, $g(k)$, by the relationship

$$\overline{g(k)} = \sum_{\text{all } k} g(k)P(k) \tag{25}$$

We note that if $g(k) = H(k) + I(k)$ — that is, if $g(k)$ equals the sum of two other functions — then

$$\begin{aligned}\overline{g(k)} &= \overline{[H(k) + I(k)]} = \sum_k [H(k) + I(k)]P(k) \\ &= \sum H(k)P(k) + \sum I(k)P(k) = \overline{H(k)} + \overline{I(k)}\end{aligned} \tag{26}$$

In other words, the average of the sum of two functions equals the sum of their averages. It should also be noted that the average of anything is simply a number. Hence, the average of an average is simply the average itself, for

$$\bar{\bar{g}} = \sum_k \bar{g}P(k) = \bar{g}\sum_k P(k) = \bar{g}$$

If the outcome of the experiment is not discrete, then we must use a distribution function $f(x)$. In the same fashion as above, we are led to define the average of any function $g(x)$ by

$$\boxed{\overline{g(x)} = \int_{x_1}^{x_2} g(x)f(x)\,dx} \tag{27}$$

where the limits on the integral extend over all possible values of x.

To illustrate the concept of averages for continuous distributions, consider the distribution function

$$f(x) = \sqrt{\frac{\beta}{\pi}}\, e^{-\beta x^2} \qquad (-\infty \leq x \leq +\infty) \tag{28}$$

where β is some constant. This distribution function turns out to be very important in statistical mechanics, and is known as the **Gaussian distribution function.** The evaluation of various integrals involving the Gaussian function is discussed in Appendix A. Thus, for example, the average value of x is given by

$$\bar{x} = \sqrt{\frac{\beta}{\pi}} \int_{-\infty}^{\infty} xe^{-\beta x^2}\,dx$$

This integral vanishes because the integrand is an odd function of x, hence $\bar{x} = 0$. On the other hand the average value of x^2 is

$$\overline{x^2} = \sqrt{\frac{\beta}{\pi}} \int_{-\infty}^{\infty} x^2 e^{-\beta x^2}\,dx \equiv 2\sqrt{\frac{\beta}{\pi}}\, G_2$$

The value of G_2 is shown in Appendix A to be equal to $(1/4\beta)\sqrt{\pi/\beta}$, hence $\overline{x^2} = 1/(2\beta)$. Thus the larger the value of β, the smaller the value of x^2. To appreciate what this means it is useful to plot $f(x)$ for two values of β. This is done in Figure 10. Note that the area under both curves must be equal [for the total probability equals $\int_{-\infty}^{\infty} f(x)\,dx = 1$, for any value of β]. The smaller the value of β, the "fatter" is the curve and the larger the value of $\overline{x^2}$. A moment's thought shows that this is reasonable. For if β is small [so $f(x)$ is broad], then there is a good chance of observing large values of x^2; hence its average value should be large.

As another example, consider the function $|x|$, which equals the magnitude of x. Its average value is

$$\overline{|x|} = \sqrt{\frac{\beta}{\pi}} \int_{-\infty}^{\infty} |x|e^{-\beta x^2}\,dx = 2\sqrt{\frac{\beta}{\pi}} \int_{0}^{\infty} xe^{-\beta x^2}\,dx = 2\sqrt{\frac{\beta}{\pi}}\, G_1$$

where G_1 is again evaluated in Appendix A. Since $G_1 = 1/(2\beta)$, we find that $\overline{|x|} = 1/\sqrt{\pi\beta}$. This value should be compared with the value of $(\overline{x^2})^{1/2} = (2\beta)^{-1/2}$. We see that they depend on β in the same manner, but are not quite equal to each other.

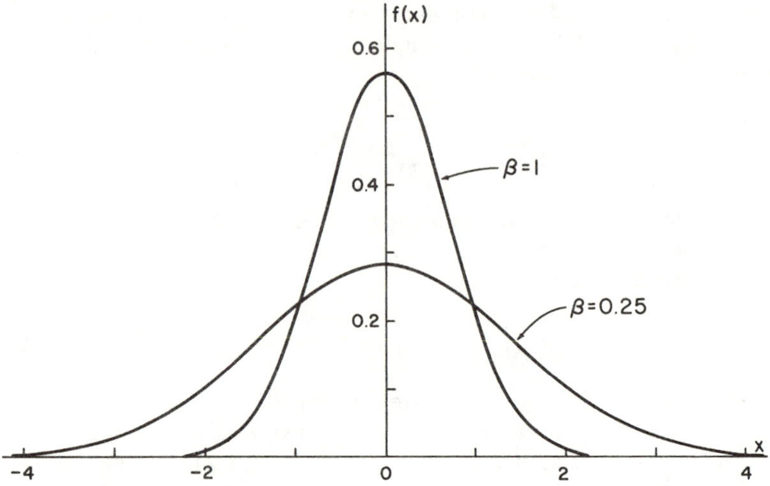

Figure 10. The distribution function $f(x) = \sqrt{\beta/\pi}\, e^{-\beta x^2}$ for two values of β.

If the Gaussian distribution is of the form

$$f(x) = \sqrt{\frac{\beta}{\pi}}\, e^{-\beta(x-x_0)^2} \tag{29}$$

then it has a maximum at $x = x_0$, rather than $x = 0$ as in the case of (28). This distribution function is illustrated in Figure 11. It is not difficult to

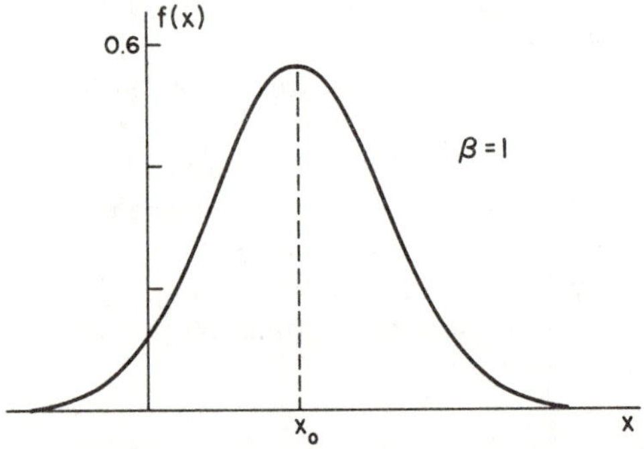

Figure 11. The "shifted" Gaussian distribution function $\sqrt{\beta/\pi}\, e^{-\beta(x-x_0)^2}$.

show that now $\bar{x} = x_0$ rather than zero. To illustrate how averages are evaluated with the shifted Gaussian distribution function, we shall now consider $\overline{x^2}$.

The average of x^2 is given by (setting $y = x - x_0$)

$$\overline{x^2} = \sqrt{\frac{\beta}{\pi}} \int_{-\infty}^{\infty} x^2 e^{-\beta(x-x_0)^2}\, dx = \sqrt{\frac{\beta}{\pi}} \int_{-\infty}^{\infty} (y + x_0)^2 e^{-\beta y^2}\, dy$$

$$= \sqrt{\frac{\beta}{\pi}} \int_{-\infty}^{\infty} (y^2 + 2yx_0 + x_0^2) e^{-\beta y^2} = \sqrt{\frac{\beta}{\pi}} (2G_2 + 0 + x_0^2\, 2G_0) = \frac{1}{2\beta} + x_0^2$$

Thus the evaluation is somewhat more involved for the "shifted" Gaussian, Equation (29), but it can be evaluated in terms of the basic Gaussian integrals G_n.

The average value of a quantity is not the only information that is of interest. Another point of interest is how much the observed value of the quantity is likely to differ in any *particular* experiment from its average value. For example, in the case of the single die, the average value of the number that turns up is 3.5. Obviously, the die *never* has a number equal to its average value (this is often the situation in discrete cases). How much of a difference from 3.5 can we expect to observe in any particular experiment? If this difference is large, then clearly the average value is not a good indication of the value that will be observed in a particular experiment (and this is usually the question of most interest). On the other hand, if this difference is small, then the average value is very significant, even for a single experiment.

If the value k occurs in an experiment, then its difference from the average value is $(k - \bar{k})$. If we compute the average value of $(k - \bar{k})$, we might expect that this would indicate the average deviation of k from its average value. However, using definition (25), we find that

$$\overline{(k - \bar{k})} = \sum (k - \bar{k}) P(k) = \sum k P(k) - \bar{k} \sum P(k) = \bar{k} - \bar{k} = 0$$

(Note that $\bar{k}$ is simply a number, and hence can be taken outside the summation sign.) Thus, the average value of $(k - \bar{k})$ vanishes, so it cannot be used to determine the average deviation of a particular measurement from its average value. The trouble is that $(k - \bar{k})$ can be both positive and negative. To obtain a quantity that does not have this property we simply need to square $(k - \bar{k})$ — that is, use $(k - \bar{k})^2$. The average of this quantity also indicates the deviation from the mean, and hence is just as acceptable. If we compute this average, we find

$$\overline{(k - \bar{k})^2} = \overline{(k^2 - 2k\bar{k} + \bar{k}^2)} = \overline{k^2} - 2\bar{k}\bar{k} + \bar{k}^2 \qquad (30)$$

$$= \overline{k^2} - (\bar{k})^2 = \sigma(k)^2 = \text{var}\,(k)$$

where we have used the results in Equation (27). This quantity is called the **variance** or **dispersion** of the quantity k. In the case of a continuous variable, the variance is defined similarly:

$$\boxed{\text{var}\,(x) = \int_{x_1}^{x_2} (x - \bar{x})^2 f(x)\, dx} \qquad (31)$$

The square root of the variance $\sigma(k)$ is called the **standard deviation:**

$$\boxed{\sigma(k) = \left[\,\overline{(k - \bar{k})^2}\,\right]^{1/2}} \tag{32}$$

To illustrate these concepts we shall consider several simple examples. In the case of rolling the single die, we find that the variance equals

$$\overline{k^2} - \bar{k}^2 = (\tfrac{1}{6})(1 + 4 + 9 + \cdots + 36) - (\tfrac{7}{2})^2 = \tfrac{35}{12}$$

Thus, the standard deviation in this case is more than 1.7. In this case the standard deviation is an appreciable fraction of the average value, $\frac{7}{2} = 3.5$. Therefore we can expect that, in any single roll of the die, the value that results can deviate considerably from the average value. As a second example we consider the case of the Gaussian distribution $f(x) = \sqrt{\beta/\pi}\, e^{-\beta x^2}$, Equation (28). We have already shown that in this case

$$\overline{x^2} = \frac{1}{2\beta} \qquad \text{and} \qquad \bar{x} = 0$$

Thus the variance of x is $1/(2\beta)$, and the standard deviation is $1/\sqrt{2\beta}$. We see that if β is large, then the standard deviation is small. In this case we can expect that any particular experiment will yield a value of x near the average value $\bar{x} = 0$. However, if β is small, then the standard deviation becomes large, and the value observed in a particular experiment may differ greatly from the average value. This should also be apparent from Figure 10.

It is instructive to note the analogy between the statistical quantities above and certain geometric and mechanical quantities. Thus, the distance between the points (x, y, z) and (x_0, y_0, z_0) in geometry is

$$[(x - x_0)^2 + (y - y_0)^2 + (z - z_0)^2]^{1/2}$$

We note that this is analogous to the standard deviation

$$\sigma = \left[\sum (k - \bar{k})^2 P_k\right]^{1/2}$$

Roughly speaking, we can think of the standard deviation as the 'distance'' a particular measurement can be expected to be found from the average value. For another analogy, consider particles of mass m_i located along the x axis at x_i. In mechanics, the center of mass is located at the point:

$$\bar{x} = \frac{\sum x_i m_i}{\sum m_i}$$

We note that this is analogous to the average value

$$\bar{k} = \frac{\sum k P_k}{\sum P_k} = \sum k P_k$$

where the probability is analogous to the masses m_i. Finally, in mechanics the moment of inertia of the above system about its center of mass is defined as

$$I = \sum (x_i - \bar{x})^2 m_i$$

which is analogous with the variance. Just as the center of mass and the moment of inertia do not determine *all* the mechanical properties of a mechanical system, neither do the average and variance determine *all* the statistical properties of an experiment. Nonetheless, most of the *important* properties of a system *are* determined by its center of mass and moment of inertia. Similarly, most of the important statistical features of an experiment are determined by the average value and variance. These concepts are illustrated by the two distribution functions in Figures 12 and 13. In both of these cases the average value $\bar{x}$ is the same. Note that if a metal sheet were cut out in the form of the area under $f(x)$, it would balance on a

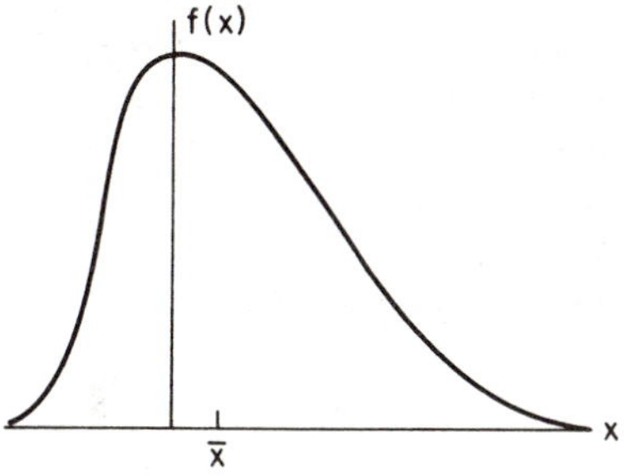

Figure 12

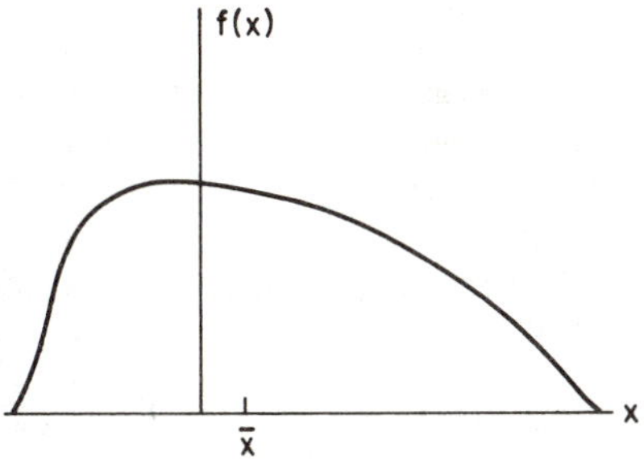

Figure 13

knife edge at the point $\bar{x}$ (the center of mass). It is clear from these figures that the probability of observing a value x near $\bar{x}$ is greater in the case of Figure 12 than Figure 13. The latter figure is more "spread out" (or "fatter"). This means that $f(x)$ in Figure 12 has a smaller variance (or dispersion) than $f(x)$ in Figure 13. The Gaussian distributions in Figures 10 and 11 have different average values but the same variance for the same value of β.

It should be noted that $\bar{x}$ and var (x) often are not dimensionless quantities, for they have the same dimensions as x and x^2, respectively. Thus, if x is a length, var (x) has the dimensions of $(\text{length})^2$. Hence if one changes units (e.g., from centimeter to meter) the numerical value of $\bar{x}$ and var (x) also changes. We see, therefore, that in these cases the numerical value of var (x) has no absolute meaning, so that a statement like "the var (x) is large" really does not mean anything. The point is that the var (x), or the dispersion $\sigma(x)$, must be compared with something else that has the same dimensions if we wish to make a meaningful statement about its magnitude. One useful dimensionless quantity is the *normalized average value* of x

$$\text{nav}(x) \equiv \frac{\bar{x}}{\sigma}$$

The numerical value of nav (x) is independent of the units that are used to measure x. If nav $(x) \ll 1$, then the value of $\bar{x}$ is not a very significant quantity — in the sense that one will often observe values of x that differ by a large percentage of $\bar{x}$ from the value $x = \bar{x}$. If, however, nav $(x) \gg 1$, then the value of x will differ by only a small percentage of $\bar{x}$ from the value $x = \bar{x}$.

A final point to note about distribution functions is that certain events are **most probable.** In the case of a continuous distribution function $f(x)$, the most probable value of x (which we denote by x_m) occurs *in the region* in which $f(x)$ is a maximum. Thus in Figures 12 and 13 the most probable values of x occur *in the region* around $x = x_m = 0$ (not *at* $x = 0$, for the probability of this *point* is zero). In the case of the Gaussian distribution shown in Figure 11 the most probable value is $x_m = x_0$. We see that the most probable value may or may not equal the average value. It should be reasonably clear from these examples that the most probable value represents the approximate value of x that is most likely to occur in *any single experiment.*

In general, x_m is given by the condition

$$\boxed{\left(\frac{df(x)}{dx}\right)_{x_m} = 0} \tag{33}$$

[that is, the maxima of $f(x)$]. The value x_m is also frequently referred to as the **mode** *of the distribution* function. If $f(x)$ has *two* maxima, it is frequently referred to as a *bimodal* distribution function. The most probable value is the mode with the largest maxima. An example of such a bimodal distribution is given in the problems.

To summarize, the average value gives information concerning the values in a large number of experiments. The variance gives a measure of how much these values are "spread out" around the average. Finally the most probable value (or the mode) represents the most likely value to occur in any single experiment.

5. *THE STATISTICAL CONCEPT OF UNCERTAINTY*

In the present section we shall examine a concept that is of considerable importance in many statistical situations and that arises in an important way in statistical mechanics. We shall call this concept "uncertainty" (a name frequently used in the field of information theory). The concept of uncertainty can be illustrated by a few simple examples. First consider two simple experiments. Case I consists in rolling a true die. The simple events in this case are the numbers $i = 1, 2, 3, \ldots, 6$, and their probabilities are $P_i = \frac{1}{6}$. Case II involves rolling a die that is not true. Assume that the probabilities in this case are $P_1 = P_2 = P_3 = P_4 = P_5 = \frac{1}{8}$, $P_6 = \frac{3}{8}$. Now, if you were allowed to select one of these dice for purposes of gambling, it should be obvious that you would select die II (and bet on number six!). The reason for choosing die II is that you are *less uncertain* about the outcome in that case than you are in case I. Indeed, if you wanted to be "dead certain" about the outcome of such an experiment, you would look for (or make!) a third die for which $P_6 = 1$, $P_i = 0$ $(i \neq 6)$. An experiment in case III clearly involves *no uncertainty*, because number six will always turn up. Let us list these results:

Case	*Probabilities*	
I	$P_i = \frac{1}{6}$ $(i = 1, \ldots, 6)$	Most uncertain of these cases
II	$P_i = \frac{1}{8}$ $(i \neq 6)$, $P_6 = \frac{3}{8}$	Less uncertain than case I
III	$P_i = 0$ $(i \neq 6)$, $P_6 = 1$	Zero uncertainty
IV	$P_i = \frac{1}{8}$ $(i \neq 3)$, $P_3 = \frac{3}{8}$	Same uncertainty as case II

Here we have added case IV, where a die is essentially the same as die II except that now number three (rather than number six) turns up with the probability $\frac{3}{8}$. Clearly the uncertainty does not depend on *which* event has a certain probability, but only on the values of *all* the probabilities. Thus the uncertainty in case IV is the same as in case II.

Consider one more point. What happens if we take die I and die III and roll them together? This, of course, represents an entirely new experiment. There are now 36 simple events (i.e., points in the sample space) represented by the pair of numbers (i, j), where i refers to die I and j to die III. Since the dice are *independent* of one another, the probabilities of these events are $P_{ij} = P_i(\text{I})P_j(\text{III})$.

What uncertainty should we assign to this new case? In the present example the answer is fairly clear. Since die III always turns up the number six, the only uncertainty is due to the other die. Hence, the uncertainty of this combination is simply equal to the uncertainty of case I. Note that we would also have arrived at this answer if we had simply *added* the uncertainty of case I and case III — that is, uncertainty (I, III) = uncertainty (I) + uncertainty (III) — because the uncertainty of III is zero. Would this idea of adding uncertainties make sense if we rolled die I and die II together? It is certainly reasonable to say that this case is at least as uncertain as either case I or case II, simply because there are more possible events. Since adding the uncertainty of I to that of II would yield a number at least as large as either one separately, we see that the *addition* also makes sense in this case. Notice that whether we should add uncertainties is not something to be proved — it is simply a choice that agrees with certain qualitative features we usually ascribe to uncertainty.

Let us now list these properties of "uncertainty," and then try to find a *quantitative* way to measure the uncertainty of an experiment. We have seen that the following properties are reasonable:

1. The uncertainty of an experiment consisting of two *independent* experiments (e.g., rolling two dice) equals the *sum* of their individual uncertainties.
2. The uncertainty of an experiment depends on the probabilities of *all* the events P_i. Thus, the uncertainty is some *average* property of the experiment.
3. The maximum uncertainty of an experiment occurs if all the probabilities are equal (as in case I). If one event has probability one, then the uncertainty is zero (as in case III).
4. The uncertainty should depend on the various P_i in a symmetric fashion (e.g., the uncertainty of case II should be the same as that of case IV).

Let us denote the uncertainty of an experiment, which has n simple events, by $H(P_i, P_2, \ldots, P_n)$. We shall *not* show how this function $H(P_1, \ldots, P_n)$ can be determined from the required properties 1 to 4, because this would involve too much analysis. Instead, we shall write down an expression for $H(P_1, \ldots, P_n)$ and show that it has all the desired properties of uncertainty. We now *define* the **uncertainty** of an experiment that has n simple events to be

$$\boxed{H(P_1, \ldots, P_n) = -\sum_{i=1}^{n} P_i \ln P_i} \tag{34}$$

Probably the most mysterious part of this expression is the logarithm term† — why should it be present? We need this term in order to satisfy condition 1. To illustrate this fact, consider again the experiment involving the rolling of two dice, say I and II. The probability of a pair of numbers (i, j) — i referring to die I and j to die II — is $P_{ij} = P_i(\mathrm{I})P_j(\mathrm{II})$. The reason is that the dice are *independent* — an important fact to note! What is the uncertainty of this experiment? According to Equation (34) it is

$$H(P_{ij}) = -\sum_{i,j=1}^{6} P_{ij} \ln P_{ij} = -\sum P_i(\mathrm{I})P_j(\mathrm{II}) \ln [P_i(\mathrm{I})P_j(\mathrm{II})]$$

†We shall use the natural logarithm for later purposes. The use of other logarithms simply involves a multiplicative constant on the right side of Equation (34).

Now the crucial property of the logarithm is that

$$\ln [P_i(\mathrm{I})P_j(\mathrm{II})] = \ln P_i(\mathrm{I}) + \ln P_j(\mathrm{II})$$

Hence

$$H(P_{ij}) = -\sum_{i,j} P_i(\mathrm{I})P_j(\mathrm{II}) \ln P_i(\mathrm{I}) - \sum_{i,j} P_i(\mathrm{I})P_j(\mathrm{II}) \ln P_j(\mathrm{II}).$$

Now we can sum over j in the first summation, and over i in the second summation, and use the fact that

$$\sum_{j=1}^{6} P_j(\mathrm{II}) = 1, \qquad \sum_{i=1}^{6} P_i(\mathrm{I}) = 1$$

to obtain

$$H(P_{ij}) = -\sum_{i=1}^{6} P_i(\mathrm{I}) \ln P_i(\mathrm{I}) - \sum_{j=1}^{6} P_j(\mathrm{II}) \ln P_j(\mathrm{II})$$

But the last two sums are just the uncertainty of experiments I and II, respectively. Thus, we have shown that expression (34) makes the uncertainty of the compound experiment equal to the sum of the uncertainties of the two experiments. It should be clear that we needed the logarithm term to accomplish this result.

It is easy to see that property 2 is also satisfied by (34). In fact, we could write $H = -\overline{\ln (P)}$ (i.e., the average of $\ln P$). The minus sign in (34) simply makes H a positive quantity — for, since all $P_i \leq 1$, all of the terms $\ln P_i$ are negative. Likewise, it is easy to see that property 4 is satisfied, because expression (34) is symmetric in the variables P_i. The fact that (34) satisfied property (3) is not entirely obvious, but nonetheless it is true. One can make it seem plausible by considering examples. Thus, for experiment I,

$$H(\mathrm{I}) = -\sum_{i=1}^{6} \tfrac{1}{6} \ln \tfrac{1}{6} = -6 \times \tfrac{1}{6} \ln \tfrac{1}{6} = -\ln \tfrac{1}{6} = \ln 6$$

Similarly, for experiment II,

$$H(\mathrm{II}) = -5 \times \tfrac{1}{8} \ln \tfrac{1}{8} - \tfrac{3}{8} \ln \tfrac{3}{8} = \tfrac{5}{8} \ln 8 + \tfrac{3}{8} \ln \tfrac{8}{3}$$

The claim is that $H(\mathrm{I}) > H(\mathrm{II})$ — that is, the uncertainty is greatest in experiment I. In fact, $H(\mathrm{I}) = 1.79$, whereas $H(\mathrm{II}) = \frac{5}{8} \times 2.08 + \frac{3}{8} \times 0.98 = 1.67$. For any other selection of values for $P_1, \ldots, P_6$ $\left(\sum P_i = 1\right)$ one can show that H must be less than $H(\mathrm{I}) = 1.79$. In the case when some $P_i = 0$, we interpret the term $P_i \ln P_i$ as the limit of $P_i \ln P_i$ as $P_i \to 0$, and one has $\lim_{P_i \to 0} P_i \ln P_i = 0$. Therefore, for experiment III,

$$H(\mathrm{III}) = -(0 + 0 + 0 + 0 + 0 + 1 \ln 1) = 0$$

since $\ln 1 = 0$.

Historically, the function H was first considered by Ludwig Boltzmann in 1872, in order to prove certain results in the kinetic theory of gases. Since he was interested in a situation involving continuous variables (the velocities of gas atoms), he did not use the form $-\sum P_i \ln P_i$, but rather its analog for continuous variables. If $f(x)\,dx$ is the probability of an event's occurring in the region dx about x, then we can use Boltzmann's expression for the uncertainty

$$\boxed{H = -\int f(x) \ln f(x)\, dx = -\overline{\ln f(x)}} \tag{35}$$

where the integral extends over all allowable values of x. This expression has nearly all the desired properties of uncertainty that we discussed above.† For example, if we have two variables, say x and y, we then should obviously take

$$\boxed{H = -\iint f(x, y) \ln f(x, y)\, dx\, dy} \tag{36}$$

Now, if $f(x, y) = g(x)\,h(y)$ (that is, the events along x are *independent* of those along y), then it is not difficult to show that

$$H = -\int g(x) \ln g(x)\, dx - \int h(y) \ln h(y)\, dy \tag{37}$$

so the uncertainties are again added together [see whether you can prove (37) — remember that $g(x)$ and $h(x)$ are normalized to unity].

The principal use of these expressions for the uncertainty is to enable us to affix some number to the *difference* in the uncertainty of two experiments. Consider, for example, the case where a particle is located somewhere between $2 \geq x \geq 0$. If we assume that all regions are equally probable, then $f(x) = \frac{1}{2}$ [for $\int_0^2 f(x)\, dx = 1$]. Then $H = \ln 2$. Now if the particle is allowed to move anywhere in the region $5 \geq x \geq 0$, we assume that $f(x) = \frac{1}{5}$, in which case $H = \ln 5$. The fact that it can be anywhere in the larger region has increased the uncertainty from $\ln 2$ to $\ln 5$. It is reasonable to say that we lose some *information* about the location of the particle when we allow it to meander around in the larger region. The amount of information we *lose* is $\ln 5 - \ln 2 = \ln \frac{5}{2}$. This is an example of a change in the uncertainty due to a real physical change in the experiment. The uncertainty in an experiment may also change if we are given some new information about the possible events. If we are told that some event A occurs in the experiment, the new probabilities for events i are the conditional probabilities $P(i \mid A)$, so the new uncertainty is

$$H(A) = -\sum_i P(i \mid A) \ln P(i \mid A) \tag{38}$$

†Some of the differences are noted in the problems.

[$H(A)$: the uncertainty, given the fact that A occurs]. For example, the uncertainty in the case of rolling a true die is ln 6 (*assuming* equal probabilities). However, if someone comes along and tells us the die rolls only numbers up to four, he has given us some *information*. The new uncertainty is

$$H(i \leq 4) = -\sum_i P(i \mid i \leq 4) \ln P(i \mid i \leq 4)$$

and since

$$P(i \mid i \leq 4) = \frac{P[i\mathrm{U}(i \leq 4)]}{P(i \leq 4)} = \frac{\frac{1}{6}}{\frac{4}{6}} = \frac{1}{4}$$

(provided $i \leq 4$), one finds $H(i \leq 4) = \ln 4$. The new uncertainty is, of course, smaller than before, and we can say that our informant gave us $\ln 6 - \ln 4 = \ln \frac{3}{2}$ units of information. Notice that when we use a priori probability, the probabilities of events — and hence the uncertainty — are based on our reasonable assumption *using all available information*. If we acquire new information about the experiment, we assume new probabilities, and hence we obtain a different uncertainty. In this case the uncertainty changes, not because the experiment is different, but simply because we have new information on which to estimate the probabilities. Usually, however, we shall be interested in the change in uncertainty effected by some real physical change in the experimental situation (such as letting the particle move around in a larger region, or gas atoms move around in a bigger volume).

ESSENTIAL POINTS

1. The probability of simple events i must satisfy the conditions

$$P_i \geq 0, \qquad \sum_i P_i = 1$$

The values of P_i are frequently assigned on the basis of a reasonable assumption (yielding a priori probabilities). The validity of this assumption rests on the agreement of the theory with the experimentally determined values of $P_i = \lim_{N \to \infty} n_i(N)/N$ (a posteriori probabilities).

2. *Definitions:*

(a) The probability of a compound event

$$P(A) = \sum_{i \subset A} P_i$$

(b) The intersect of A and $B \equiv AB =$ all events that A and B have in common.

(c) Conditional probability $P(A \mid B) \equiv P(AB)/P(B) =$ probability that A occurs given the fact that B occurs.

(d) A and B are independent if (and only if) $P(AB) = P(A)P(B)$ — or what is the same thing $P(A \mid B) = P(A)$.

3. For continuous events we define a distribution function $f(x, y, \ldots)$ of one or more variables by

$$P(x, dx; y, dy; \ldots) \equiv f(x, y, \ldots)\, dx\, dy \cdots$$

$=$ probability of the occurrence of the event in the infinitesimal region $(dx, dy, \ldots)$ about the values $(x, y, \ldots)$.

4. *Definitions:*

(a) Average value:

$$\overline{k} = \sum_k kP(k); \qquad \overline{g(x)} = \int g(x)f(x)\, dx$$

(b) Variance:

$$\text{var}\,(x) = \int (x - \bar{x})^2 f(x)\, dx = \overline{(x - \bar{x})^2} = \overline{x^2} - (\bar{x})^2$$

(c) Most probable value (mode):

$$\left[\frac{df(x)}{dx}\right]_{x_m} = 0 \qquad [\text{maximum of } f(x)]$$

5. The uncertainty of a statistical situation can be measured by

$$H = -\sum P_i \ln P_i \qquad \text{or} \qquad H = -\int f(x) \ln f(x)\, dx$$

where the summation (or integral) extends over all possible events. The uncertainty of an experiment consisting of two *independent* parts equals the *sum* of the uncertainties of the two parts.

PROBLEMS

1. The 2 of clubs is lost from a deck of cards. The experiment consists in drawing one card from this deck.

(a) Determine the probabilities $P(X)$, $X = D, H, C, S$, of the events of drawing a diamond, heart, club, or spade. State your assumptions, and then prove your result.

(b) Determine the probabilities $P(i)$, $i = 2, 3, \ldots, 14$, of the events of drawing a 2, 3, . . . , ace.

(c) Determine for what values of X and i, $P(iX) = P(X)P(i)$.

(d) What can you conclude from (c)?

2. Consider the following experiment: Among the digits 1, 2, 3, 4, first one is chosen, and then a second selection is made from the remaining three digits. The simple events are these *ordered* pairs of numbers [for example, (1, 3), (3, 1), (3, 4), and so on].

(a) How many points are there in the sample space? Make a reasonable assumption and determine the probability of each event.

(b) Consider the compound events

A: the first digit is odd; B: the first digit is even

Determine $P(A)$ and $P(B)$. State the basic principle used.

(c) Determine $P(A \cup B)$ and $P(AB)$. Show that

$$P(A \cup B) = P(A) + P(B) - P(AB)$$

Determine whether the events A and B are independent. State why this result is reasonable.

(d) Consider the compound events

C: the last digit is even; D: the last digit is odd

What is the value of $P(C)$ and $P(D)$?

(e) Draw a sample space with all the points (do not label them) and indicate all the compound events A, B, C, and D.

(f) Determine $P(C \cup A)$ and $P(AC)$, and show that

$$P(A \cup C) = P(A) + P(C) - P(AC)$$

(g) What is the probability that, if the first digit is known to be odd, the last digit will be even? To determine this, first express the statement above in terms of the appropriate conditional probability, and then use Equation (9) to determine its value. Are the two events independent? Why?

(h) Let E be the compound event: the sum of the two digits is even. Determine $P(E)$, $P(A \cup E)$, and $P(AE)$. Show that

$$P(A \cup E) = P(A) + P(E) - P(AE)$$

(i) What is the probability that, if the sum of the digits is known to be even, the first digit is odd? Follow the procedure in part (g). Are the two events independent? Why?

3. A factory has three machines A, B, and C which respectively produce 25, 35, and 40 per cent of the resistors made by the company. Of this output, these machines respectively produce 5, 4, and 2 per cent defective resistors. A resistor is selected at random from the output of this company and is found to be defective. What is the probability it was produced by the machine A, B, and C? To answer this question, first determine the possible results of picking a resistor from the output (events) and obtain the sample space. Relate the information above to the probability of certain compound events, or to points in the sample space. Draw the sample space and all relevant compound events. Use these results to analyze this problem.

4. Consider the experiment shown in Figure 5. Assume that the distribution function is of the form

$$f(x) = \frac{CL^2}{(x^2 + L^2)^{3/2}}$$

where C is a constant and L is the distance between the screen and the gas container.

(a) Determine the constant C. [Hint: $L^2dx/(x^2 + L^2)^{3/2} = d(x/(x^2 + L^2)^{1/2})$.]

(b) If $L = 2m$, what is the probability that an atom will strike the screen between the points $x = -1$ cm and $+1$ cm?

(c) What is the probability that an atom will strike in the region $x \leq 1$ m?

(d) If a total of 10^6 atoms strike the screen, what is the probable number of atoms that would be found in the region dx of $x = 3$ m? At the point $x = -1$ m?

5. In the manufacturing of an electronic instrument, it is found that the probability $P(n)$ that an instrument has n defects within six months is

$$P(0) = 0.1, \quad P(1) = 0.4, \quad P(2) = 0.25,$$
$$P(3) = 0.15, \quad P(4) = 0.08, \quad P(5) = 0.02$$

(a) What is the average number of defects of all the instruments in the first six months?

(b) If you bought an instrument, what is the most likely number of defects it would have in the first six months?

(c) If the manufacturer must repair the instrument, which of the numbers — (a) or (b) — is significant in determining his expense?

6. The speed s of cars on a road is found to be given by the distribution function

$$f(s) = As \exp\left(\frac{-s}{s_0}\right) \qquad (0 \leq s \leq \infty)$$

where A and s_0 are constants.

(a) Determine A in terms of s_0.

(b) A radar unit can differentiate only between speeds that differ by small amounts Δs. In the region of what speed s_m is it most likely to find a particular car?

(c) What is the probability of the radar unit's actually finding a car in this region?

(d) What is the average speed of the cars?

(e) Assume that the number of accidents that a car has is proportional to its speed — say, bs per month (where b is some constant). What is the average number of accidents per month on this road, assuming that N cars use it?

7. A man who enjoys throwing darts at a vertical pole placed against a wall finds from long experience that the probability of a dart's hitting the wall at a distance x from the center of the pole (where $x = 0$) is well represented by the distribution function

$$f(x) = \frac{\lambda}{\sqrt{\pi}} e^{-\lambda^2(x-x_0)^2} \qquad (-\infty \leq x \leq \infty)$$

where λ and x_0 both depend on the day of the week.

(a) What is the average distance $\bar{x}$ by which he misses the target in terms of λ and x_0?

(b) What is the dispersion of his shots?

(c) If he throws 200 darts, what is the probable *number* of darts that hit between x_1 and $x_1 + dx_1$?

(d) He finds that λ and x_0 vary with the days as follows:

	λ (*inches*)$^{-1}$	x_0 (*inches*)
Monday	$\frac{1}{2}$	2.0
Wednesday	$\frac{1}{3}$	−1.0
Friday	$\frac{1}{4}$	0.0

On what day does he hit most consistently near his average position?

(e) On what day is he most likely to hit the target? If the target is 0.1 inch wide, what is the probability (approximately)?

8. (a) Can the distribution function [Equation (20)]

$$f(x, y) = \frac{1}{\pi} \frac{L^2}{(L^2 + x^2 + y^2)^2}$$

be written in the form $f_1(x)f_2(y)$ (that is, the product of two other functions that depend on x and y separately)? If so, determine $f_1(x)f_2(y)$, taking into account the normalization of both distribution functions. Explain the significance of this result, using Equations (11) and (19).

(b) Repeat part (a) for the distribution function [Equation (23)]

$$F(r, \theta) = \frac{rL^2}{\pi(L^2 + r^2)^2}$$

[where, of course, the product would now be $F_1(r)F_2(\theta)$]. Note that these two distributions $f(x, y)$ and $F(r, \theta)$ both refer to the same physical experiment.

9. Consider the distribution function for the velocity of a particle $f(\mathbf{v}) = (c/\pi)^{3/2} e^{-c\mathbf{v}^2}$, where c is a constant and each velocity component can range from $-\infty$ to $+\infty$.

(a) Using the results of Appendix A, show that $f(\mathbf{v})$ is normalized to unity. Note that this involves three integrals that can be evaluated separately.

(b) Determine the probability that $v_x \geq 0$, $v_y \geq 0$, and $v_z \leq 0$ (i.e., simultaneously). Explain its dependency on the constant c.

(c) Obtain an expression for the probability that the x component of $\mathbf{v}$ is in the range dv_x of v_x, regardless of the values of v_y and v_z. Explain your reasoning [along the lines of Equation (21)].

10. An interesting example of a bimodal distribution function is provided by Old Faithful, the famous geyser in Yellowstone National Park. If we let t represent the time between the eruptions of the geyser (in minutes), then the probability $f(t)dt$ that the next eruption occurs in an interval dt of t can be reasonably represented by the sum of two Gaussians

$$f(t) = 0.34\sqrt{\frac{\beta_1}{\pi}}\,e^{-\beta_1(t-51)^2} + 0.66\sqrt{\frac{\beta_2}{\pi}}\,e^{-\beta_2(t-74)^2}$$

where $\beta_1 = 0.021$ and $\beta_2 = 0.017$.

(a) Determine to the nearest minute the time when $f(t)$ is a maximum [that is, the two modes of $f(t)$]. Which is the most probable value of $f(t)$? Note that when one Gaussian is a maximum, the other Gaussian is essentially zero.

(b) What is the value of $\bar{t}$? To obtain this, integrate over $-\infty \leq t \leq \infty$. The result is essentially the same as integrating over $0 \leq t \leq \infty$ because $f(t) \simeq 0$ for negative values of t.

(c) Using the same range of integration as in (b), determine the standard deviation, $\sigma(t)$ [by way of comparison, if $f(t) = (\sqrt{\beta_2}/\pi)e^{-\beta_2(t-74)^2}$ then $\sigma(t) \simeq 5.4$.] What percentage of $\bar{t}$ is $\sigma(t)$? Is Old Faithful faithful?

11. Consider the three distribution functions shown below (all are normalized to unity). Decide the *relative* order of magnitude of the variances var (I), var (II), and var (III) and arrange them in increasing values. Do the same for the three uncertainties H(I), H(II), and H(III). Note that the ordering is not the same in the two cases. Explain your reasoning.

12. A black and a white marble are each dropped into one of four boxes (or two different atoms move about in a container that is mentally divided into quarters). Determine the possible number of events. Using a reasonable assumption, determine the uncertainty. What is the uncertainty if the marbles must be in separate boxes? Explain the reason for the difference in H.

13. (a) Determine the uncertainty H for the Gaussian distribution function

$$f(x) = \frac{\lambda}{\sqrt{\pi}}\,e^{-\lambda^2(x-x_0)^2} \qquad (-\infty \leq x \leq \infty)$$

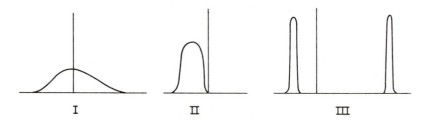

showing how it is related to var (x). Note that H is independent of x_0. Why does this make sense?

(b) Determine H if

$$f(x) = ce^{-cx} \qquad (0 \leq x \leq \infty)$$

showing how it is related to $\bar{x}$. Prove that, if we make the var (x) the same for the two distributions in parts (a) and (b) (by relating c and λ), then the uncertainty is greatest in the case of the Gaussian distribution function (this can be shown to be a general property of the Gaussian distribution).

14. The uncertainty in the case of continuous distributions, as defined by $H = -\int f(x) \ln f(x) \, dx$, differs in several respects from the case of discrete events. Because $f(x)$ can be larger than unity, $\ln f(x)$ can be positive so that H may be negative (in contrast to the case of discrete events) and may even approach $-\infty$. One interpretation of such values is that continuous distributions can contain an infinite amount of information, and the fact that H is negative simply means that the uncertainty is less than when H is positive. Thus the situation $H = 0$ does not signify that there is no uncertainty, as it does in the discrete case. Only the *differences* in the values of H are of any significance.

(a) To illustrate this point, determine the values of λ and C in the distributions of Problem 13 which yield $H = 0$.

(b) Determine how the properties of the quantity $\tilde{H} = -\int f(x) \ln [f(x)/f(x_m)] \, dx$ differ from those of H (here x_m is the most probable value of x). From this determine the largest negative value which H can have for a given distribution function $f(x)$.

2

Energy

1. INTRODUCTION

In the previous chapter we discussed one of the two basic ingredients of statistical mechanics — probability — together with related statistical concepts such as averages, variance, and uncertainty. The second basic ingredient of statistical mechanics is, as its name implies, mechanics. More specifically, it is the concept of energy. Classical thermodynamics, of course, also deals with energy, but only as it arises in *macroscopic* interactions between systems. Thus it is concerned with the transfer of energy in various macroscopic processes such as changing the volume of a system (and thereby doing $p\,dV$ work) or heating it with a flame. A fundamental observation of classical thermodynamics is that the work required to go from one equilibrium state to another, in an adiabatic process, is independent of the methods used to perform the work (i.e., the work in an adiabatic process depends only on the end states). This observation, which implies the existence of the state function U (the internal energy), we now interpret as "simply" a macroscopic example of the conservation of energy. Its importance, however, lies in the fact that it is a completely general law (true for all systems) that does not depend on any microscopic picture we may dream up about what is going on inside the system.

In statistical mechanics, on the other hand, in which we consider the microscopic nature of these systems, we must deal with microscopic (or mechanical) forms of energy as it appears in the motion and in the interaction between the atoms of the system (the kinetic and potential energy). The energy exchange observed in macroscopic (thermodynamic) processes is, of course, related to the microscopic form of energy — and one of the objectives of statistical mechanics is to make this relationship clear. In the present chapter we shall discuss some of the most important aspects of energy and how it can be described in terms of various component parts. We shall begin with an elementary classical description of the energy of a point particle and then discuss some of the forms of energy that arise because of the *interaction* between particles. Some *models* (simplified pictures) of atoms, molecules, gases, and solids will illustrate the various types of interactions between particles. Finally, we shall consider some of the important modifications of these classical models that are required by quantum mechanics.

2. KINETIC AND POTENTIAL ENERGY

To begin with, let us consider a particle moving in space as it is described in classical mechanics. The position of the particle $\mathbf{r}(t) = x(t)\mathbf{i} + y(t)\mathbf{j} + z(t)\mathbf{k}$ [where $(\mathbf{i}, \mathbf{j}, \mathbf{k})$ are the unit vectors in the (x, y, z) directions] generally varies in time and has a velocity

$$\mathbf{v} = v_x\mathbf{i} + v_y\mathbf{j} + v_z\mathbf{k} \equiv \frac{dx}{dt}\mathbf{i} + \frac{dy}{dt}\mathbf{j} + \frac{dz}{dt}\mathbf{k} = \frac{d\mathbf{r}}{dt}$$

If the particle has a mass m, it is said to have a **kinetic energy** (sometimes called the *translational energy*)

$$\epsilon_{\text{kin}} = \tfrac{1}{2}m\mathbf{v}^2 \equiv \tfrac{1}{2}m(v_x^2 + v_y^2 + v_z^2) \tag{1}$$

If a force $\mathbf{F}$ acts on the particle, its velocity will change according to Newton's law

$$\mathbf{F} = m\frac{d\mathbf{v}}{dt} \tag{2}$$

If no force acts on the particle, then it follows from Equation (2) that $\mathbf{v}$ is independent of the time, and consequently the kinetic energy is also a constant. We can also see this from the fact that

$$\frac{d\epsilon_{\text{kin}}}{dt} = m\left(v_x\frac{dv_x}{dt} + v_y\frac{dv_y}{dt} + v_z\frac{dv_z}{dt}\right) = m\mathbf{v}\cdot\frac{d\mathbf{v}}{dt} = \mathbf{v}\cdot\mathbf{F} \tag{3}$$

which vanishes if $\mathbf{F}$ vanishes.

Now the force that acts on the particle generally depends on the position of the particle $\mathbf{r}$ and sometimes also on the velocity (e.g., frictional forces, or the force on a charged particle moving in a magnetic field). If the force depends only on $\mathbf{r}$ and if it can be expressed in terms of the gradient of some function $\phi(\mathbf{r})$ through the relationship

$$\mathbf{F}(\mathbf{r}) = -\nabla\phi \equiv -\left(\frac{\partial\phi}{\partial x}\mathbf{i} + \frac{\partial\phi}{\partial y}\mathbf{j} + \frac{\partial\phi}{\partial z}\mathbf{k}\right) \tag{4}$$

then we say that the force field $\mathbf{F}(\mathbf{r})$ is *conservative*. In what follows we shall consider only such conservative forces. The function $\phi(\mathbf{r})$ in Equation (4) is called the **potential energy** of the particle (at the point $\mathbf{r}$). We can see the importance of this function by again considering Equation (3). We have, using Equation (4),

$$\frac{d\epsilon_{\text{kin}}}{dt} = \mathbf{v}\cdot\mathbf{F} = -\mathbf{v}\cdot\nabla\phi = -\left(\frac{dx}{dt}\frac{\partial\phi}{\partial x} + \frac{dy}{dt}\frac{\partial\phi}{\partial y} + \frac{dz}{dt}\frac{\partial\phi}{\partial z}\right) \equiv -\frac{d\phi}{dt} \tag{5}$$

Now Equation (5) can be written

$$\frac{d}{dt}[\epsilon_{\text{kin}} + \phi(\mathbf{r})] = 0$$

This shows that the quantity

$$\epsilon = \tfrac{1}{2}m\mathbf{v}^2 + \phi(\mathbf{r}) \tag{6}$$

is independent of time (i.e., it is constant), even when the force does not vanish. The quantity ϵ is called the (total) *energy* of the particle. It is the sum of the kinetic and potential energy. Note that, although neither the kinetic nor the potential energy is generally constant, their sum is constant. The fact that the total energy is constant is known as the **conservation of energy.**

It is important to note that the *magnitude* of the energy does not have any physical significance — only the *change* in the energy can be measured. This can be seen by the fact that, since $\mathbf{F} = -\nabla\phi$, we can add any constant to $\phi(\mathbf{r})$ and not change the force. Hence, only the difference in the potential energy at two points can be physically significant. To see what this difference represents, consider again Equation (5). We have

$$-\frac{d\phi}{dt} = \mathbf{F}\cdot\mathbf{v} \equiv \mathbf{F}\cdot\frac{d\mathbf{r}}{dt}$$

Therefore

$$-d\phi = \mathbf{F}\cdot d\mathbf{r} \tag{7}$$

The right side is simply the work done on the particle by the force $\mathbf{F}$ when the particle moves a distance $d\mathbf{r}$. If we want to move the particle, we must exert a force $-\mathbf{F}$, so we shall do the work $-\mathbf{F}\cdot d\mathbf{r} = d\phi$. Thus

$$\int_{r_1}^{r_2} d\phi \equiv \phi(\mathbf{r}_2) - \phi(r_1)$$

$$= \text{the external work that must be done to move the particle from the point } \mathbf{r}_1 \text{ to } \mathbf{r}_2 \tag{8}$$

A few examples will illustrate these points. First, consider the case of the gravitational force. This force is proportional to the mass of the particle and is nearly constant over reasonable distances. Let z be the height above the ground and mg the force (g is the gravitational constant, $g \simeq 32$ ft sec$^{-2} \simeq 980$ cm sec^{-2}). Since the force is only in the z direction,

$$F_z = -mg \qquad (F_x = F_y = 0)$$

(The minus sign is there because the force is toward the ground.) Then, according to Equation (4), the gravitational potential energy is

$$\phi = mgz + \text{constant} \tag{9}$$

The constant in ϕ has no physical significance, and it can be selected arbitrarily. The work required to raise the particle from the ground to the height z is

$$\int_0^z mg\,dz = mgz = \phi(z) - \phi(0)$$

where the last equality follows from (9) regardless of the value of the constant. We see that this agrees with Equation (8). If we set this constant equal to zero, then the total energy of the particle is, in this case,

$$\epsilon = \tfrac{1}{2}m\mathbf{v}^2 + mgz \tag{10}$$

As z decreases (the particle drops toward the ground), v must increase in such a way that ϵ remains constant. Thus, the decrease in the potential energy is compensated for by the increase in the kinetic energy.

For the second example we consider one of the most important simple mechanical systems. A particle is connected to a fixed point ($x = 0$) by a spring. If the spring obeys Hooke's law, then the force it exerts on the particle is proportional to the amount the spring is stretched from its undisturbed (equilibrium) length. Assume that the length of the spring, if left undisturbed, is x_0. Then, if x is the position of the particle, $(x - x_0)$ represents the amount the spring is stretched (or compressed), and Hooke's law says that

$$F_x = -\kappa(x - x_0)$$

where κ is called the *force constant* (see Figure 1). The force is always such as to restore the particle to the point $x = x_0$, for $F_x < 0$ if $x > x_0$, whereas

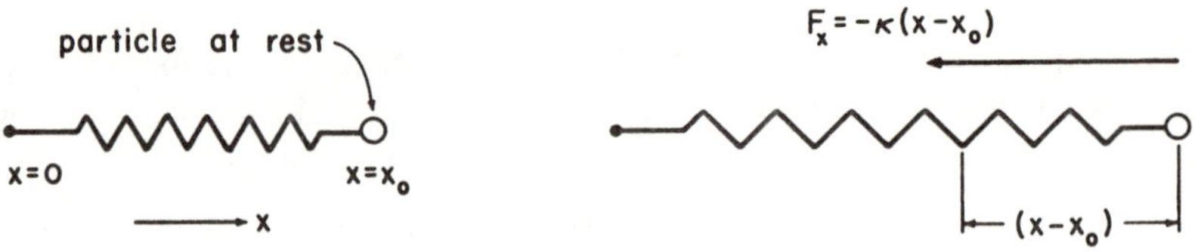

Figure 1

$F_x > 0$ if $x < x_0$. However, when the particle reaches the point x_0, it has a large velocity so that it overshoots this point. In other words, it oscillates back and forth past $x = x_0$. This system is an example of a **harmonic oscillator.** The potential energy in this case is [if we take $\phi(x_0) = 0$]

$$\phi = \tfrac{1}{2}\kappa(x - x_0)^2$$

The distinctive feature of a harmonic potential is that it is a *quadratic function of* $(x - x_0)$. The total energy is

$$\epsilon = \tfrac{1}{2}mv^2 + \tfrac{1}{2}\kappa(x - x_0)^2 \tag{11}$$

To verify that ϵ is constant in time, we first must solve Newton's equation:

$$m\frac{d^2x}{dt^2} = \kappa(x - x_0)$$

The general solution of this equation is

$$x - x_0 = A\cos\omega t + B\sin\omega t \qquad \left(\omega^2 = \frac{\kappa}{m}\right) \ (A,\ B\text{: constants})$$

as may be verified by direct substitution. The motion is periodic with a period $2\pi/\omega$, so the frequency is $\nu = (1/2\pi)(\kappa/m)^{1/2}$. It is left as an exercise to show that if this solution is substituted into Equation (11), using $v = dx/dt$, then ϵ is in fact a constant.

If $\phi(x)$ is plotted as a function of x, we obtain a figure like Figure 2. Sometimes a potential of this shape is referred to as a "potential well" or "potential valley." At the bottom of the well, where $d\phi/dx = 0$, there is no

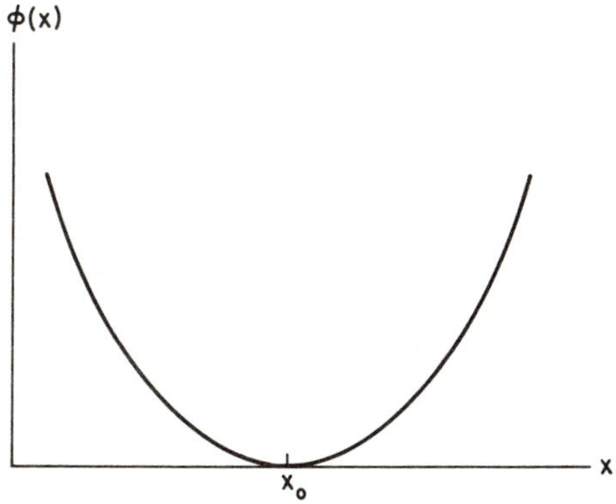

Figure 2

force on the particle. A particle placed at that point remains there, and it is therefore called the point of *mechanical equilibrium.* To the right of the equilibrium point, the force is toward the left ($F_x = -d\phi/dx$), whereas to the left of x_0 the force is directed toward the right ($d\phi/dx < 0$) — like a ball rolling in a bowl shaped like $\phi(x)$.

If a particle is bound to a point $\mathbf{r}_0$ by a harmonic force and it can move in *three* directions, then its energy is

$$\epsilon = \tfrac{1}{2}m\mathbf{v}^2 + \tfrac{1}{2}\kappa(\mathbf{r} - \mathbf{r}_0)^2 \tag{12}$$

Note that the force is always directed toward the point $\mathbf{r}_0$, for

$$\mathbf{F} = -\nabla(\tfrac{1}{2})\kappa(\mathbf{r} - \mathbf{r}_0)^2 = -\kappa(\mathbf{r} - \mathbf{r}_0)$$

and is proportional to the displacement $\mathbf{r} - \mathbf{r}_0$ (harmonic force).

Another important class of systems are those that react to an applied electric field. If two charges $\pm q$ are separated by a distance $\mathbf{r}$, we say that they have a *dipole moment* $\mathbf{p} = q\mathbf{r}$ (see Figure 3). Such situations exist, for example, in those atoms in which the electrons are not situated symmetri-

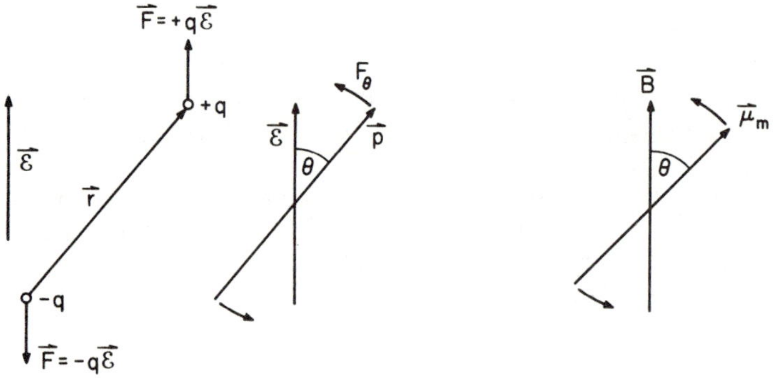

Electric case

Magnetic case

Figure 3

cally around the positively charged nucleus. When such a dipole is placed in an electric field, the force on the two charges tries to orient $\mathbf{p}$ in the direction of $\boldsymbol{\mathcal{E}}$ (see Figure 3). That is, the electric field produces a torque on the dipole (or a force in the θ direction), given by

$$F_\theta = -p\mathcal{E}\sin\theta$$

(The minus sign is because the force is in the direction of smaller θ.) Thus the potential is (since $F_\theta = -\partial\phi/\partial\theta$)

$$\phi = -p\mathcal{E}\cos\theta = -\boldsymbol{\mathcal{E}}\cdot\mathbf{p} \tag{13}$$

where we have taken $\phi(\theta = \pi/2) = 0$. The order of magnitude of the dipole moment for a molecule can be estimated using the fact that the size of a molecule is a few angstroms (1 Å = 10^{-8} cm). If an electron and a proton (charge: 4.8×10^{-10} esu; 1.6×10^{-19} coulomb) are separated by one angstrom, the resulting dipole moment is

$$p = 4.8 \times 10^{-10} \times 10^{-8} = 4.8 \times 10^{-18} \text{ (esu-cm)}$$

or

$$p = 1.6 \times 10^{-19} \times 10^{-10} = 1.6 \times 10^{-29} \text{ (coulomb-m)}$$

Tabulated values of p for various molecules are usually in units of 10^{-18} (esu-cm), which is called a "debye." Some values of p are given in Table 1 in both esu units and rationalized mks units.

Table 1

	Permanent Dipole Moment			*Polarizability*
Molecule	p (× 10^{+18} *esu-cm*)	p (× 10^{29} *coul-m*)	α (Å^3)	α (× 10^{40} *coul-m²/volt*)
H_2	0.00	0.00	0.79	0.88
CO	0.12	0.04	1.99	2.21
HCl	1.03	0.34	2.63	2.92
H_2O	1.86	0.62	1.48	1.64
NH_3	1.46	0.50	2.26	2.51

If a molecule is placed in an electric field $\mathcal{E}$ it also often has a dipole moment *induced* by the field,

$$\mathbf{p}_{\text{ind}} = \alpha \mathcal{E}$$

where α is called the polarizability of the molecule. The last two columns in Table 1 give the value of the polarizability of these molecules. The potential energy of the molecule in the field $\boldsymbol{\mathcal{E}}$ is now given by

$$\phi = \boldsymbol{\mathcal{E}} \cdot \mathbf{p}_{\text{per}} - \tfrac{1}{2}\alpha \mathcal{E}^2 \tag{14}$$

A final important example of the energy of a molecule due to an external force field is the case of a magnetic moment $\boldsymbol{\mu}_m$ in a magnetic field with flux density **B**. Molecules may have a magnetic moment for a number of reasons, but the simplest to visualize is the magnetic moment that results from the electrons circling the nucleus of an atom. A circling electron behaves like a little current loop, having a magnetic moment equal to the current times the area of the loop. The direction of $\boldsymbol{\mu}_m$ is normal to the loop. When this loop is placed in a magnetic field **B**, a torque acts on it which tries to align $\boldsymbol{\mu}_m$ parallel to **B** (see Figure 3). The magnitude of this torque is $-\boldsymbol{\mu}_m B \sin\theta$, so the potential energy

$$\phi(\theta) = -\mu_m B \cos\theta = -\boldsymbol{\mu}_m \cdot \mathbf{B} \tag{15}$$

In mks units μ_m is in units of amp-m^2 and B is in webers/m^2 (1 weber/m^2 = 10^4 gauss). Typical values of μ_m are of the order of 10^{-23} amp-m^2. Thus for an oxygen molecule, $\mu_m \simeq 3 \times 10^{-23}$ amp-m^2, whereas a fancy molecule like gadolinium sulfate [$Gd_2(SO_4)_3 \cdot 8H_2O$] has a molecular magnetic moment of 7.2×10^{-23} amp-m^2. It should be noted that the energy in the electric case, Equation (13), is very similar to the magnetic case, Equation (15). The magnetic case also has, in general, an induced magnetic moment $\alpha_m B$, which is called the *diamagnetic term*. We shall consider, however, only the case of the permanent magnetic moment, which is called the *paramagnetic term*.

3. INTERACTION ENERGY

In the last section we discussed only the energy of a single particle due to its motion (kinetic energy) and externally applied forces. In statistical mechanics, in which we are concerned with macroscopic systems, we must deal with a very large number of particles (i.e., electrons, protons, and neutrons). Such particles possess energy, not only because of their motion and external forces, but also because the particles exert forces on one another. These interaction forces are, of course, what give rise to atoms, molecules, liquids, and solids. Although it is not particularly accurate to describe some of these interactions in classical (i.e., nonquantum-mechanical) terms, we shall nonetheless do so in this section. Thus we shall discuss *classical models* of various macroscopic systems, which are quite useful as a first approximation but have certain limitations. Some of these limitations due to quantum-mechanical effects will be taken up in Section 5.

Before considering the interaction between particles, we should first define the kinetic and potential energy of a collection of noninteracting particles (they do not exist, of course, but it is often useful to consider such a model). If a system contains N particles, the total kinetic energy is defined simply as the sum of the individual kinetic energies of the particles:

$$E_{\text{kin}} = \tfrac{1}{2} m_1 \mathbf{v}_1^2 + \tfrac{1}{2} m_2 \mathbf{v}_2^2 + \cdots + \tfrac{1}{2} m_N \mathbf{v}_N^2 = \sum_{i=1}^{N} \tfrac{1}{2} m_i \mathbf{v}_i^2 \tag{16}$$

where m_i, $\mathbf{v}_i$) is the mass and velocity of particle i. Moreover, if these particles are in an *external* (conservative) force field, such as an electric field or gravity, then the total potential energy due to these forces is again the sum of their individual potential energies:

$$\phi_{\text{ext}} = \sum_{i=1}^{N} \phi(\mathbf{r}_i) \tag{17}$$

Note that, just as E_{kin} depends on the value of each velocity $\mathbf{v}_i$, ϕ_{ext} depends on the position of each particle $\mathbf{r}_i$.

Now we turn to the question of the interactions between particles and between molecules. Just as we can describe conservative *external* forces in terms of a potential energy $\phi(\mathbf{r}_i)$, we can frequently represent the *intermolecular forces* by a potential energy $\Phi(r_{ij})$. In this case

$$r_{ij} = |\mathbf{r}_i - \mathbf{r}_j| = [(x_i - x_j)^2 + (y_i - y_j)^2 + (z_i - z_j)^2]^{1/2}$$

is the distance between the molecules, and the force on particle i (due to particle j) is given by

$$\mathbf{F}_i = -\frac{\partial\Phi(r_{ij})}{\partial x_i}\mathbf{i} - \frac{\partial\Phi(r_{ij})}{\partial y_i}\mathbf{j} - \frac{\partial\Phi(r_{ij})}{\partial z_i}\mathbf{k}$$

If the potential depends only on the magnitude of the separation between the molecules, then

$$\mathbf{F}_i = -\frac{d\Phi(r_{ij})}{dr_{ij}}\frac{\mathbf{r}_{ij}}{r_{ij}} \qquad (\mathbf{r}_{ij} = \mathbf{r}_i - \mathbf{r}_j) \tag{18}$$

and the force is along their connecting line, as illustrated in Figure 4. Note that $\mathbf{F}_i = -\mathbf{F}_j$, so the forces are equal but in opposite directions.

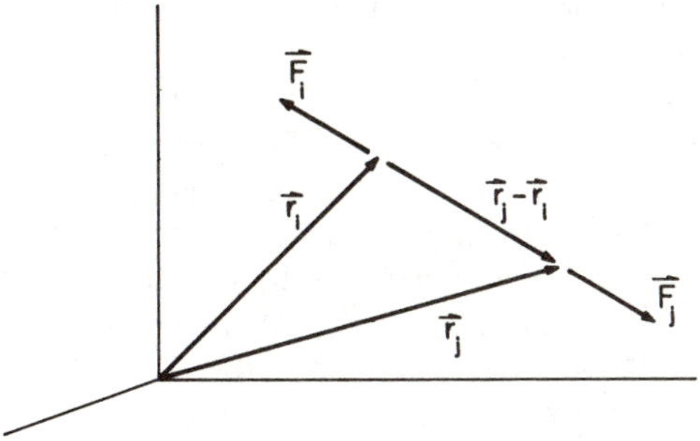

Figure 4. Intermolecular forces

The intermolecular forces can be conveniently divided into repulsive forces and attractive forces. These repulsive forces may originate in electrical forces, nuclear forces, or certain quantum-mechanical effects. For example, when two neutral atoms come very close together, the electrons around each atom begin to overlap. Not only do the electrons repel each other because of Coulomb forces, but they also have another (much stronger) aversion to being in the same region of space. This aversion is a quantum-mechanical effect (the exclusion principle), which need not concern us at present. The essential feature of these repulsive forces is that they increase very rapidly as the separation distance decreases. In fact, a useful model of this repulsive force is obtained by considering the atoms to be simply *hard spheres* (small billiard balls) with a fixed radius, say r_{HS}. In this model the force between two atoms becomes infinite when they come into contact (when $|\mathbf{r}_1 - \mathbf{r}_2| = 2r_{HS}$), so the potential $\Phi(r)$ has an infinite negative slope at this point. This is illustrated in Figure 5. A somewhat more realistic model takes $\Phi(r) = C/r^n$, where n is a large number. The

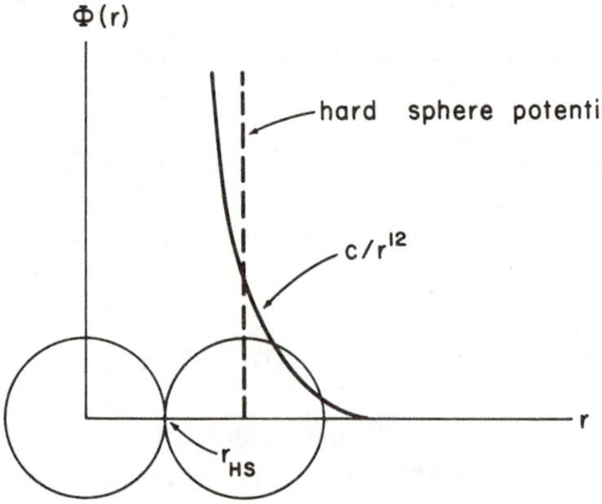

Figure 5. Models of the repulsive potentials between molecules as a function of their separation distance $r = |\mathbf{r}_1 - \mathbf{r}_2|$

fact that n is a large number means that the force increases rapidly as the separation distance decreases. One of the most commonly used potentials is proportional to r^{-12}, also shown in Figure 5. In this case, if r decreases by a factor of $\frac{1}{2}$, the potential energy increases by a factor of 4,096 ($= 2^{12}$), and hence the potential is quite similar to a hard-sphere potential.

Perhaps the most common attractive force between particles is the Coulomb force between two particles of charge q_1 and q_2:

$$\mathbf{F}_1 = \frac{q_1 q_2}{4\pi\epsilon_0} \frac{\mathbf{r}_{12}}{r_{12}^3} = -\mathbf{F}_2$$

where $\mathbf{r}_{12} = \mathbf{r}_1 - \mathbf{r}_2$. The force is attractive if q_1 and q_2 have opposite signs, and repulsive otherwise. The potential energy is, according to definition (18),

$$\Phi(r_{ij}) = \frac{q_1 q_2}{4\pi\epsilon_0} \frac{1}{r_{ij}} \tag{19}$$

where the q's are in coulombs and $\epsilon_0 = 8.85 \times 10^{-12}$ coul^2/newton-m^2 (rationalized mks units). This potential decreases very slowly as r_{ij} is increased. The attractive force between an electron and a proton is, of course, what gives rise to the bound electrons in atoms. A simple *classical* picture of a hydrogen atom is one in which the attractive Coulomb force on the electron $e^2/4\pi\epsilon_0 r^2$ is balanced by the centrifugal force mv^2/r due to its circular motion about the proton. Equating these two forces yields the condition

$$\tfrac{1}{2} mv^2 = \frac{e^2}{8\pi\epsilon_0 r}$$

The total energy of the electron is therefore

$$\epsilon_{\text{elec}} = \tfrac{1}{2} mv^2 - \frac{e^2}{4\pi\epsilon_0 r} = -\frac{e^2}{8\pi\epsilon_0 r} \quad (20)$$

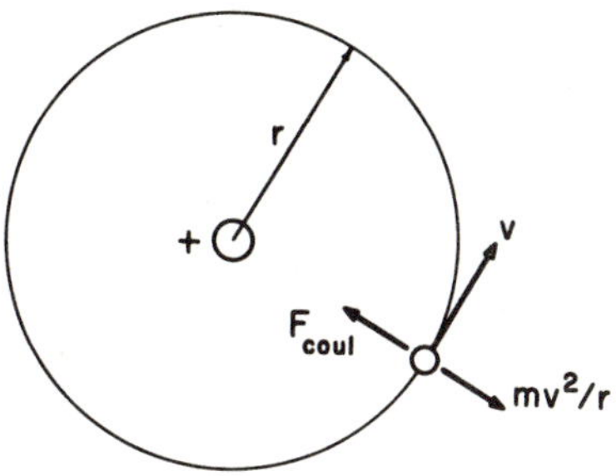

Figure 6

In this simple classical picture of a hydrogen atom, the energy of the electron can have a continuous range of values (depending on the value of r). We know, however, that this classical model of an atom is *not* in agreement with experiments, and that the energy of the electron can only have certain discrete values — a point we shall return to in Section 5.

The Coulomb force not only holds atoms together, but it is largely responsible for the formation of certain molecules and solids. If an electron that is normally bound to one atom transfers to another atom, the resulting positive and negative ions are strongly attracted to each other, in what is called an ionic bond. This attractive Coulomb force is largely responsible for the formation of such molecules as NaCl (Na^+, Cl^-), KCl, and CsF. The energy required to separate these molecules into individual ions is very large (on an atomic scale) — about 5 ev per molecule (1 electron volt = 1.6×10^{-19} joules).

A much weaker, but important, attractive force between molecules is the so-called *attractive van der Waals force*. Although this force also arises from Coulomb interactions, it is due to the attraction between electric dipoles (which were described in the last section). It is not difficult to see that two electric dipoles can produce a weak attractive force on one another, even though they are electrically neutral. Consider Figure 7(a), which shows two asymmetric molecules whose dipole moments are oriented in opposite

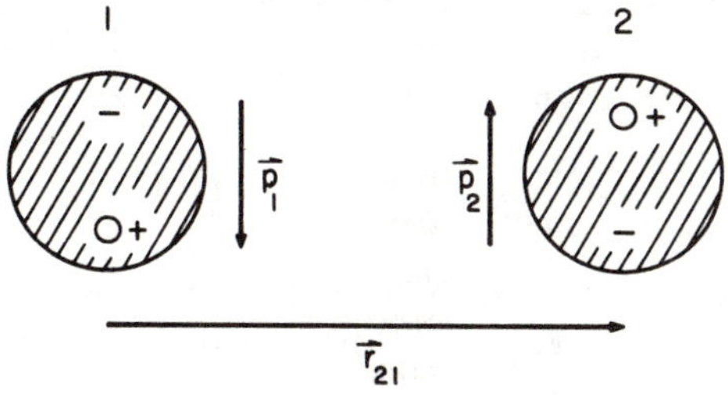

Figure 7(a)

directions. The shaded area represents the average location of the electrons. In Figure 7(b) we see that the electric field due to $\mathbf{p}_1$ does two things to the second dipole. First, it exerts a relatively strong component of force along the direction of $\mathbf{p}_2$, which tends to orient $\mathbf{p}_2$ in the way shown and also to *induce* a dipole moment in that direction. Second, a small force tends to push *both* the positive and negative charges toward the left. Therefore, dipole 1 attracts dipole 2, and similarly dipole 2 attracts dipole 1, and this yields the van der Waals attractive force. As might be expected, this force decreases more rapidly than the Coulomb force as r_{12} increases. It turns out that the potential energy in this case decreases as r^{-6} and depends upon the magnitude of p_1 and p_2 (and the polarizabilities α_1, α_2). This potential energy, therefore, has the form

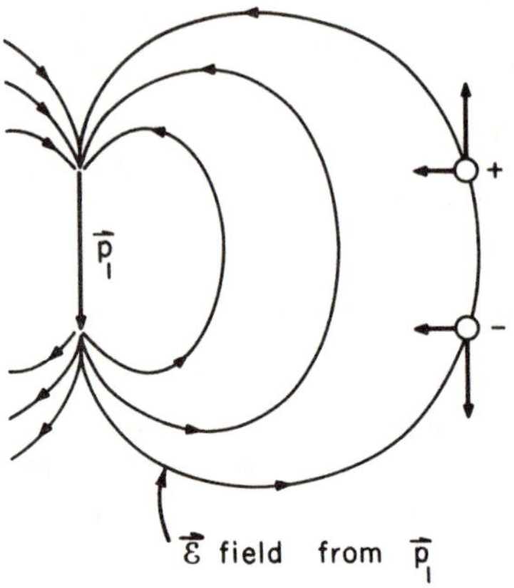

Figure 7(b)

$$\Phi(r) = -\frac{A}{r^6} \qquad (A: \text{positive constant}) \tag{21}$$

The minus sign is due to the fact that the resulting force is attractive [as you can verify by using Equation (18)].

Now, if we add the attractive van der Waals potential energy to the repulsive potential energy $\Phi(r) = C/r^{12}$, we obtain an intermolecular potential energy of the form

$$\Phi(r) = \epsilon_0\left[\left(\frac{r_0}{r}\right)^{12} - 2\left(\frac{r_0}{r}\right)^{6}\right] \tag{22}$$

where ϵ_0 and r_0 are constants that depend on the type of molecules. This form of the intermolecular potential energy is called the **Lennard-Jones potential.** It is illustrated in Figure 8. The force between the molecules is along their connecting line and has a magnitude given by

$$-\frac{d\Phi}{dr} = \frac{12\epsilon_0}{r_0}\left[\left(\frac{r_0}{r}\right)^{13} - \left(\frac{r_0}{r}\right)^{7}\right]$$

Thus the force vanishes when the molecules are separated by a distance $r = r_0$. The potential energy at this distance is $-\epsilon_0$ [using Equation (22)]. Several values of ϵ_0 and r_0 are given in Table 2 for the Lennard-Jones potential between *similar molecules* (e.g., between one He and another He atom,

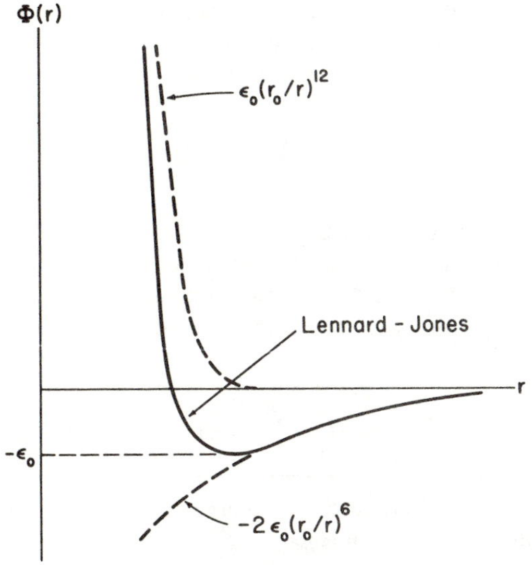

Figure 8

or between one O_2 *molecule* and another O_2 *molecule*). Note that the binding energy between these molecules (the energy required to pull them apart, which is ϵ_0) is only about 10^{-14} ergs $\simeq 10^{-2}$ electron volts (ev). This is much smaller than the 5 ev required to separate a molecule such as NaCl. In particular it should be noted that the attractive force between He atoms is *very* weak, and one might expect, therefore, that these atoms would not readily form molecules (He_2) — to say nothing about forming a liquid or solid.

Table 2

Molecule	ϵ_0 ($\times 10^{15}$ *ergs*)	r_0 (Å $= 10^{-8}$ *cm*)
He	1.42	2.87
Ne	4.83	3.12
Ar	16.7	3.82
Kr	24.8	4.04
Xe	30.7	4.60
O_2*	16.4	3.88
CO*	13.9	4.23
HCl*	50.1	3.72

*These values depend slightly on the temperature and are given for the temperature range of 100 to 300° K.

The total energy of a macroscopic system is the sum of the total kinetic energy E_{kin}, plus the total potential energy due to external forces ϕ_{ext}, plus the total potential energy due to interactions within the system. The latter potential energy is given by

$$\frac{1}{2}\sum_{i\neq j}^{N}\Phi(r_{ij})$$

where i does not equal j (a particle does not interact with itself). The factor of $\frac{1}{2}$ is due to the fact that the sum contains identical terms — such as $\Phi(r_{12})$ and $\Phi(r_{21})$ — which should only be counted once. Using the expressions for E_{kin}, Equation (16), and ϕ_{ext}, Equation (17), we finally obtain for the **total energy**

$$\boxed{E = \sum_{i=1}^{N}\frac{1}{2}m_i\mathbf{v}_i^2 + \sum_{i=1}^{N}\phi(\mathbf{r}_i) + \frac{1}{2}\sum_{i\neq j}^{N}\Phi(r_{ij})} \tag{23}$$

The importance of E is that, although there are forces between particles, the total energy *does not change in time.*

The fact that the total energy is the *sum* of various contributions leads to one other important observation. Consider two macroscopic systems (i.e., containing a *large number of particles*) which we label A and B. The total energy E_{A+B} of the *composite* system $(A + B)$ is given by Equation (23), where the sums are over all particles in both A *and* B. This expression can also be written in the form

$$E_{A+B} = E_A + E_B + E_{\text{int}} \tag{24}$$

where E_A (or E_B) is given by Equation (23) except that the sums are only over the particles in A (or in B). The energy that remains in E_{A+B} we call the interaction energy between the (macroscopic) systems.

$$E_{\text{int}} = \sum_{i\subset A}\sum_{j\subset B}\Phi(r_{ij}) \tag{25}$$

In these sums $i \subset A$ (read: i contained in A) means that i is to be summed only over the particles in system A, and similarly j is to be summed only over the particles in system B. Now if A and B are macroscopic systems, then the interaction energy E_{int} is small compared with E_A and E_B:

$$E_A \gg E_{\text{int}}, \qquad E_B \gg E_{\text{int}} \tag{26}$$

The reason for this is that most of the terms in Equation (25) are zero unless r_{ij} is very small (say, 3 to 4×10^{-8} cm), which means that the particles i and j must be in a small region of space near the partition between systems A and B (the "interaction region" in Figure 9). Hence, for macroscopic systems, Equation (24) can be approximated by

$$\boxed{E_{A+B} \simeq E_A + E_B} \tag{27}$$

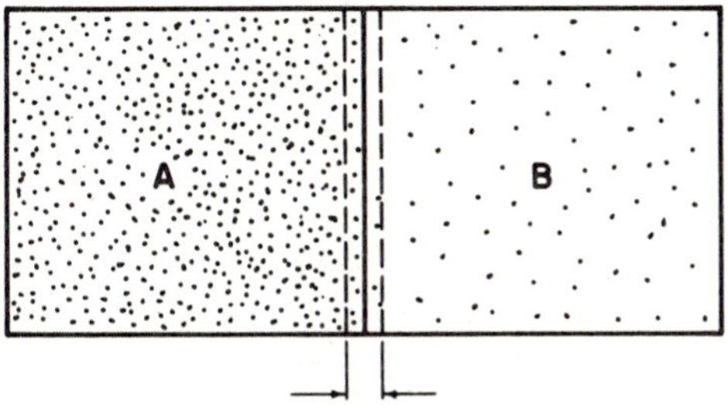

Figure 9

and we can speak of the energy of systems A and B (E_A and E_B). It is important to note, however, that although E_{A+B} is constant in time, the energies E_A and E_B can change because of the forces between the systems. A schematic picture of the possible change in E_A and E_B is shown in Figure 10.

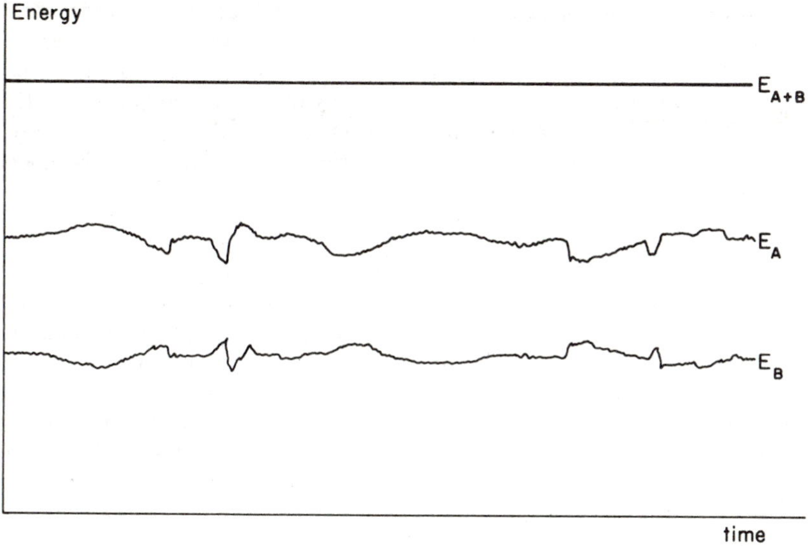

Figure 10

4. *MODELS OF MACROSCOPIC SYSTEMS*

One of the troubles with the real world is that it is complicated. Being only human, we would certainly like to ignore some of this complexity — the "unnecessary" complexity. This leads us, rather naturally, to attempt to make a simplified picture (model) of a real system that at the same time retains the system's essential physical features.

What are the "essential physical features"? This is difficult to say. Unfortunately there is no general rule for sifting out these important features — the search usually reduces to one of trial and error. In the present section we shall consider some of the models of macroscopic systems that have resulted from a long history of trials. These models naturally have their limitations, for they will all be **classical models** but they are very useful for developing a "physical feeling" about various systems. The quantum-mechanical corrections, which are essentially for explaining many properties at low temperatures, will be considered in Section 5. It should always be borne in mind that models are simply mental constructs; they aid in our understanding of nature, but, as Ludwig Boltzmann noted, "Our theories bear the same relationship to nature as symbols do to the entities they stand for, or as the letters of the alphabet bear to the voice, or as notes do to music." With these words of warning about the significance we should attach to models, let us consider several different examples.

Models for Gases. A gas is a system in which the molecules are not too close together, except for infrequent collisions between the molecules. A reasonable simplification is to assume that the *interaction energy between the molecules is negligible,* so that the total energy of the system is equal to the sum of the energies of the molecules. Since this simplification should represent a real gas better for very low densities, we expect that this model represents a perfect gas, so

$$\boxed{E = \sum_{i=1}^{N} \epsilon_i \qquad \text{(perfect gas)}} \tag{28}$$

where ϵ_i is the energy of the ith *molecule.* The type of perfect gas depends on the molecules involved — monatomic, diatomic, or polyatomic.

Consider first the simplest case, the monatomic molecule. The energy of an atom can be broken down into several parts. The most important of these are usually the translational (kinetic) energy of the atom $\epsilon_{\text{trans}} = \frac{1}{2}m\mathbf{v}^2$ (where m is the mass of the atom), the potential energy due to external forces $\phi(\mathbf{r})$, and the energy of the electrons bound to the nucleus ϵ_{elec}. Adding these up, we have

$$\boxed{\epsilon_i = \tfrac{1}{2}\, m_i\mathbf{v}_i^2 + \phi(\mathbf{r}_i) + \epsilon_{\text{elec}}(i) \qquad \text{(monatomic)}} \tag{29}$$

In many cases the energy of the electrons ϵ_{elec} remains constant, in which case it can be neglected (for we can always drop a constant from ϵ). If it changes, then we must use a quantum-mechanical expression for ϵ_{elec}, which will be discussed in the next section. If we substitute Equation (29) into Equation (28), we obtain the total energy of a perfect monatomic gas.

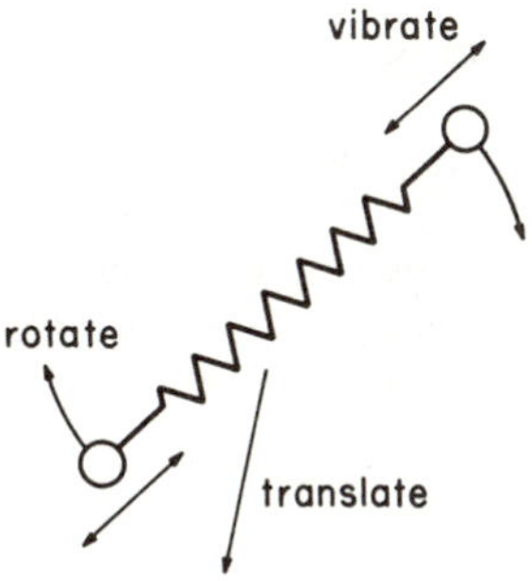

Figure 11

Next consider a diatomic molecule. As might be expected, the real situation is more complicated in this case, and we can develop several models — some better than others. Before considering any details, let us consider what two bound atoms can do. Figure 11 shows a picture of a diatomic molecule (the spring is a picturesque way of illustrating the interatomic force holding the atoms together). It is reasonably clear that the two atoms can *vibrate* along their connecting line, *rotate* about some axis, *translate* through space. The energy of such a molecule should, therefore, be the sum of the energies resulting from these motions:

$$\boxed{\epsilon = \epsilon_{\text{trans}} + \epsilon_{\text{rot}} + \epsilon_{\text{vib}} + \epsilon_{\text{pot}} \qquad \text{(diatomic)}} \tag{30}$$

The term ϵ_{pot} is the *total* potential energy due to external forces. To this we could also add the electron energy, but for simplicity we shall omit it here. Now in the *classical model*, what are the expressions for ϵ_{trans}, ϵ_{pot}, ϵ_{rot}, and ϵ_{vib}? To obtain the various types of energies we begin with the energy of the molecule:†

$$\epsilon = \tfrac{1}{2}m_a\mathbf{v}_a^2 + \tfrac{1}{2}m_b\mathbf{v}_b^2 + \Phi(r_{ab}) + \phi(\mathbf{r}_a) + \phi(\mathbf{r}_b) \tag{31}$$

where $r_{ab} = |\mathbf{r}_a - \mathbf{r}_b|$, and (a, b) refer to the two atmos. First of all, the total potential energy due to external forces is simply $\epsilon_{\text{pot}} = \phi_a(\mathbf{r}_a) + \phi_b(\mathbf{r}_b)$. The real problem is how to simplify the term $\Phi(r_{ab})$. Recall that $\Phi(r)$ depends on r in the way shown in Figure 12 (e.g., the Lennard-Jones potential).

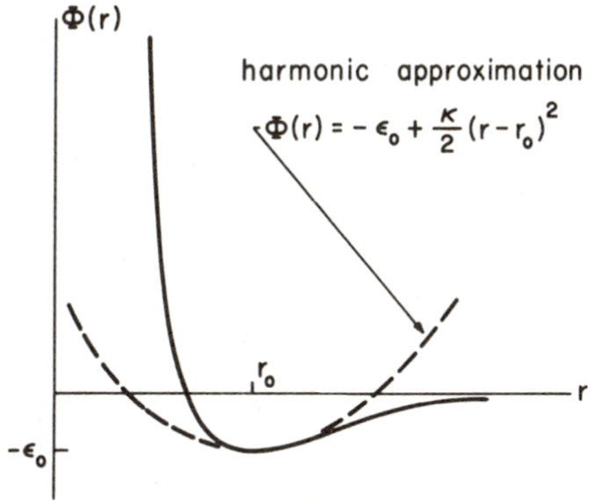

Figure 12

†Here we treat the atoms as point particles, ignoring the fact that they have a spatial extension and an electronic structure.

Since the atoms are bound together, they are separated by a distance nearly equal to r_0. Therefore, we can make the approximation

$$\Phi(r) \equiv \Phi[r_0 + (r - r_0)] \simeq \Phi(r_0) + (r - r_0)\left(\frac{d\Phi}{dr}\right)_{r_0} + \frac{1}{2}(r - r_0)^2\left(\frac{d^2\Phi}{dr^2}\right)_{r_0}$$

where we have neglected the higher order terms in the Taylor expansion. This is a good approximation if $(r - r_0)$ is not too large. Now $\Phi(r_0) = -\epsilon_0$, and $(d\Phi/dr)_{r_0} = 0$ (equilibrium point), so that

$$\Phi(r) \simeq -\epsilon_0 + \frac{1}{2}(r - r_0)^2\left(\frac{d^2\Phi}{dr^2}\right)_{r_0} \tag{32}$$

This approximation of $\Phi(r)$ will be called the **harmonic approximation,** because the resulting force $-d\Phi/dr = -(r - r_0)(d^2\Phi/dr^2)_{r_0}$ is proportional to the displacement from the equilibrium separation r_0. Note that the equivalent "force constant" is $\kappa = (d^2\Phi/dr^2)_{r_0}$ and that $-\epsilon_0$ represents the (constant) binding energy of the molecule. Hence, in the *harmonic approximation*, the classical expression for the energy, Equation (31), can be written

$$\epsilon = \frac{1}{2}m_a\mathbf{v}_a^2 + \frac{1}{2}m_b\mathbf{v}_b^2 + \frac{1}{2}\kappa(r_{ab} - r_0)^2 + \epsilon_{\text{pot}} - \epsilon_0$$

To reduce the first three terms to the form (30), we first introduce the *center of mass*

$$\mathbf{R} = \frac{m_a\mathbf{r}_a + m_b\mathbf{r}_b}{m_a + m_b}$$

which has the velocity

$$\mathbf{V} = \frac{d\mathbf{R}}{dt} = \frac{m_a\mathbf{v}_a + m_b\mathbf{v}_b}{m_a + m_b}$$

Now one can show without much trouble that

$$\frac{1}{2}m_a\mathbf{v}_a^2 + \frac{1}{2}m_b\mathbf{v}_b^2 = \frac{1}{2}(m_a + m_b)\mathbf{V}^2 + \frac{1}{2}\left(\frac{m_a m_b}{m_a + m_b}\right)(\mathbf{v}_a - \mathbf{v}_b)^2$$

The quantity $\mu = m_a m_b/(m_a + m_b)$ is referred to as the *reduced mass*, whereas $m_a + m_b = M$ is the *total mass* of the molecule. Therefore, the energy of the molecule can be written

$$\epsilon = \frac{1}{2}M\mathbf{V}^2 + \frac{1}{2}\mu\left(\frac{d\mathbf{r}_{ab}}{dt}\right)^2 + \frac{1}{2}\kappa(r_{ab} - r_0)^2 + \epsilon_{\text{pot}} - \epsilon_0 \tag{33}$$

where we have used the fact that $\mathbf{v}_a - \mathbf{v}_b = d\mathbf{r}_{ab}\,dt$. The first term $\frac{1}{2}M\,\mathbf{V}^2$ represents the translational energy of the molecule as a whole, so we set $\epsilon_{\text{trans}} = \frac{1}{2}M\,\mathbf{V}^2$. This leaves only ϵ_{vib} and ϵ_{rot} to be identified. A moment's thought (and Figure 11) shows that the vibrational motion is represented by the change in $\mathbf{r}_{ab}$ *along* the direction of $\mathbf{r}_{ab}$ (i.e., motion *along* the connecting line), whereas rotational motion is due to the change in $\mathbf{r}_{ab}$ in a direction *perpendicular* to $\mathbf{r}_{ab}$. Thus, vibrational motion changes the *magnitude* of $\mathbf{r}_{ab}$ while rotational motion changes its *direction*. To simplify matters a little, let us assume that during the rotational motion $\mathbf{r}_{ab}$ has a fixed magnitude r_0 (which is a good approximation). Then $(d\mathbf{r}_{ab}/dt)_{\text{rot}} = r_0(\boldsymbol{\omega}_1 + \boldsymbol{\omega}_2)$, where $\boldsymbol{\omega}_1$ and $\boldsymbol{\omega}_2$ are the angular velocities in two directions perpendicular to $\mathbf{r}_{ab}$. Since the change in $\mathbf{r}_{ab}$ due to vibrational motion is along $\mathbf{r}_{ab}$, we can write $(d\mathbf{r}_{ab}/dt)_{\text{vib}} = (dr_{ab}/dt)(\mathbf{r}_{ab}/r_{ab})$ — that is, the change in the magnitude times a unit vector in the direction of r_{ab}. The total rate of change of $\mathbf{r}_{ab}$ is the sum of these motions $d\mathbf{r}_{ab}/dt = (d\mathbf{r}_{ab}/dt)_{\text{rot}} + (d\mathbf{r}_{ab}/dt)_{\text{vib}}$, and since $\mathbf{r}_{ab}$, $\boldsymbol{\omega}_1$, and $\boldsymbol{\omega}_2$ are all perpendicular, we find that

$$\left(\frac{d\mathbf{r}_{ab}}{dt}\right)^2 = \left(\frac{dr_{ab}}{dt}\right)^2 + r_0^2(\omega_1^2 + \omega_2^2)$$

If we substitute this back into Equation (33), we finally obtain for the classical energy of a *diatomic molecule:*

$$\begin{aligned}\epsilon &= \frac{1}{2} M\mathbf{V}^2 + \frac{1}{2}\mu r_0^2(\omega_1^2 + \omega_2^2) + \frac{1}{2}\mu\left(\frac{dr_{ab}}{dt}\right)^2 \\ &\quad + \frac{1}{2}\kappa(r_{ab} - r_0)^2 + \phi(\mathbf{r}_a) + \phi(\mathbf{r}_0) - \epsilon_0 \\ &\equiv \epsilon_{\text{trans}} + \epsilon_{\text{rot}} + \epsilon_{\text{vib}} + \epsilon_{\text{pot}} \qquad \text{(harmonic model)}\end{aligned} \tag{34}$$

where we have identified $\frac{1}{2}\mu r_0^2(\omega_1^2 + \omega_2^2)$ as the rotational energy and set $\epsilon_{\text{vib}} = \frac{1}{2}\mu(dr_{ab}/dt)^2 + \frac{1}{2}\kappa(r_{ab} - r_0)^2$. The constant ϵ_0, which represents the binding energy of the molecule, can be ignored in most cases. If Equation (34) is substituted into (28), then we obtain the total energy of a perfect diatomic gas (harmonic model).

Another somewhat simpler model of a diatomic molecule pictures the two atoms as being connected by a rigid rod (similar to a dumbbell). This is illustrated in Figure 13. In this case there is no vibrational motion, so $\epsilon_{\text{vib}} = 0$, and one obtains

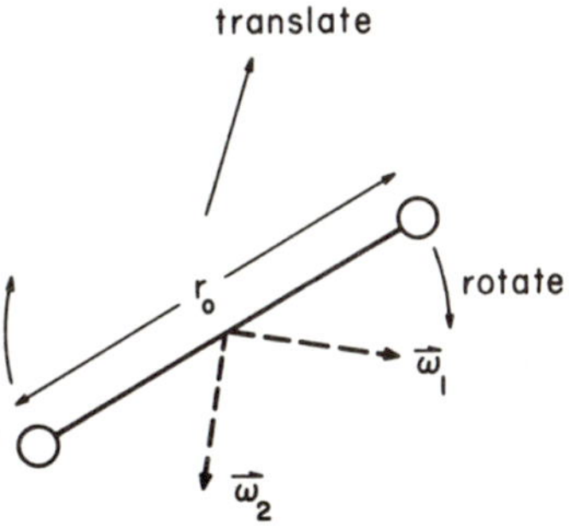

Figure 13

$$\epsilon = \tfrac{1}{2} M\mathbf{V}^2 + \tfrac{1}{2}I(\omega_1^2 + \omega_2^2) + \phi(\mathbf{r}_a) + \phi(\mathbf{r}_b) \qquad \text{(dumbbell model)} \tag{35}$$

where $I = \mu r_0^2$ is called the *moment of inertia* of the molecule. The approximate magnitude of the moment of inertia can be estimated if we know the masses of the atoms and use the fact that $r_0 \simeq 1$ to 2×10^{-8} cm. For example, for HCl the atom weights of H and Cl are approximately 1 and 35.5, respectively, and the atomic mass unit is 1.66×10^{-24} gm. So the reduced mass is $\mu = (35.5/36.5) \times 1.66 \times 10^{-24} \simeq 1.62 \times 10^{-24}$ gm. If we set $r_0 = 1.5 \times 10^{-8}$ cm, then

$$I = \mu r_0^2 \simeq 1.62 \times 2.25 \times 10^{-40} \simeq 3.64 \times 10^{-40} \text{ gm-cm}^2$$

whereas the actual value is 2.65×10^{-40} gm-cm² (indicating that $r_0 \simeq 1.28$ Å). The moments of inertia of several other diatomic molecules are listed in Table 3. Note that there is a wide variation in the value of I for different molecules.

Table 3

Molecule	$I \times 10^{40}$ *(gm-cm²)*
H_2	0.461
HF	1.34
HCl	2.65
CO	14.5
NO	16.5
Cl_2	115.0
CsCl	393.0
I_2	745.0

Models of a Solid. The essential feature that distinguishes a solid from a liquid or a gas is that the atoms of a solid are in some regular spatial arrangement. The reason they stay in such an arrangement is because they are so closely packed together that the interaction between the atoms causes each atom to "keep in its place." One such arrangement of atoms is shown in Figure 14. This is called a simple cubic lattice, and it is the structure of such solids as sodium chloride (NaCl = salt), potassium chloride (KCl), and lithium fluoride (LiF). Other solids have different arrangements. Thus Na, Li, Cs (cesium), and CsCl have a cubic lattice with an additional atom in the center of each cube (called a body-centered cubic lattice). Solids such as copper (Cu) and silver (Ag) are formed from a cubic lattice with additional atoms in the center of each face of the cube (face-centered cubic lattice). These are the relatively simple structures; the rest are more complicated. For example, a diamond has as its building block a body-centered cubic structure, except that there are carbon atoms missing at four corners. Other solids also have this structure, such as (gray) tin and silicon carbide

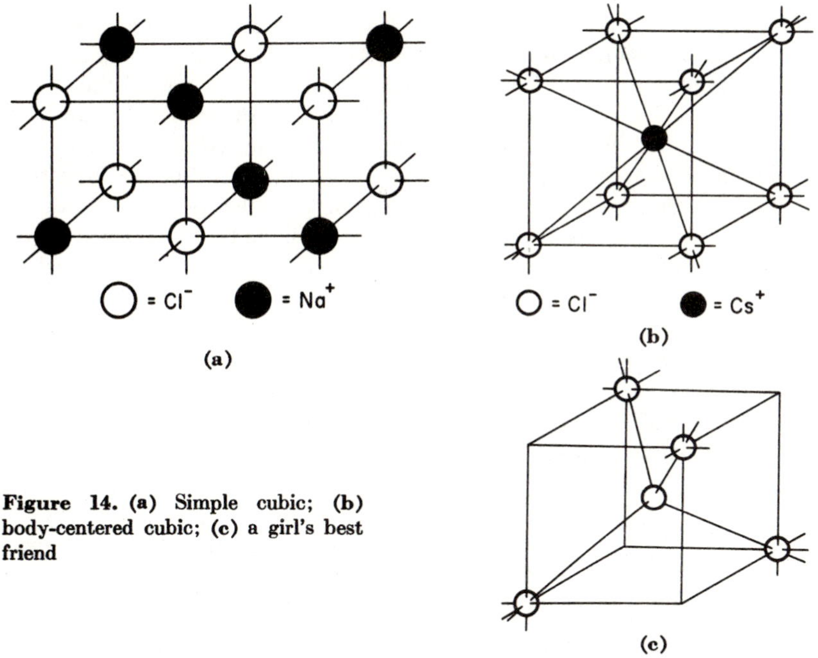

Figure 14. **(a)** Simple cubic; **(b)** body-centered cubic; **(c)** a girl's best friend

(SiC, which like diamond is noted for its hardness). These examples suffice to indicate some of the regular atomic arrangements in solids, and the fact that they can be fairly complicated.

Now the purpose of a model is, of course, to try to avoid all this complication by extracting the important contributions to the energy of the system. We might, for example, look upon a solid as simply one fantastically large molecule — so large that we can keep it from either translating or rotating in space (i.e., it sits on a table top and does not rotate). In this case the only energy it can have is due to the vibrational motion of the atoms and the possible contribution made by the electrons and outside forces. Thus the *total* energy (which we designate by a capital E) should be

$$\boxed{E = E_{\text{vib}} + E_{\text{elec}} + E_{\text{pot}}} \tag{36}$$

where E_{elec} is the total energy of the electrons and E_{pot} is the total energy due to outside forces (magnetic or electric).

Let us first consider how we can approximate the vibrational energy E_{vib}. Considering the fairly complicated structure of a solid (noted above), we might expect it to be difficult to obtain an expression for the vibrational energy. Actually, if a careful analysis is used, it *is* difficult to express E_{vib} in a useful (i.e., simple) form. However, as luck would have it, we can use an amazingly simple approximation to E_{vib} that contains most of the essential features. The idea is as follows: Each atom in the lattice can move in any of *three* directions about its equilibrium position (which is a point of minimum potential energy). If the atom does not move too far from this minimum, then the force acting on it will be proportional to its displacement from the equilibrium point. Hence its motion will be that of a *three-dimensional harmonic oscillator*, which has the energy

$$\epsilon = \tfrac{1}{2}m\mathbf{v}^2 + \tfrac{1}{2}\kappa\mathbf{r}^2$$

($\mathbf{r}$ being the displacement from the equilibrium position). Here κ is some unknown force constant, which may be different for different atoms. In this model the total vibrational energy is then taken to be the sum of the vibrational energies of each atom:

$$\boxed{E_{\text{vib}} = \sum_{i=1}^{N} \frac{1}{2} m_i\mathbf{v}_i^2 + \frac{1}{2}\kappa_i\mathbf{r}_i^2 \qquad \text{(harmonic lattice)}} \tag{37}$$

In other words, the system behaves like a collection of $3N$ *independent harmonic oscillators*. It turns out that in the classical case we do not have to estimate the values of the frequencies $\nu_i = (1/2\pi)(\kappa_i/m_i)^{1/2}$; we can leave these arbitrary. In the quantum-mechanical case (next section) we shall have to make some assumption about the values of the frequencies.

The contribution of the electrons to the energy of a solid cannot be adequately represented by any classical model. The best we can do classically is simply set E_{elec} = constant (which can then be ignored). This turns out to be very good, but the trouble is that we cannot justify this approximation by any classical argument. In metals that are good electrical conductors we know that electrons must be fairly free to move across the entire system (because this is what makes them good electrical conductors!). In other words, some electrons must not be tightly bound to the atoms — they are, so to speak, citizens of the world, belonging to no single atom but moving freely through the entire lattice. These electrons would then each have a kinetic energy of $\frac{1}{2}m_e\mathbf{v}^2$, and so they should contribute an energy

$$\boxed{E_{\text{elec}} = \sum_{\substack{\text{free}\\ \text{electrons}}} \frac{1}{2} m_e\mathbf{v}_i^2 \qquad \text{(classically)}} \tag{38}$$

where the sum is over all "free" (unbound) electrons.

Finally we must consider the contributions due to E_{pot}. There are two cases of particular interest, which we can consider together.

Simple Dielectric and Paramagnetic Systems. We noted in Section 2 that molecules may have a dipole moment $\mathbf{p}$ or a magnetic moment $\boldsymbol{\mu}_m$ and that when an electric or magnetic field is applied, they have the potential energy

$$\phi = -\mathbf{p}\cdot\boldsymbol{\mathcal{E}} \qquad \text{(electric case)}$$

$$\phi = -\boldsymbol{\mu}_m\cdot\mathbf{B} \qquad \text{(magnetic case)}$$

If a system (be it solid, liquid, or gas) is composed of such molecules, it will naturally have a corresponding potential energy due to the applied field. If the interaction energy between the dipole moments (or between the permanent magnetic moments) is very small compared with their potential energy due to the applied field, then we can obtain a simple model for such systems. These systems will be referred to as simple dielectric and paramagnetic materials in the electric and magnetic case, respectively. They are pictured as a collection of dipole or magnetic moments oriented in random directions as illustrated in Figure 15. When an external field (electric or magnetic) is applied to such a system, it acquires potential energy given by

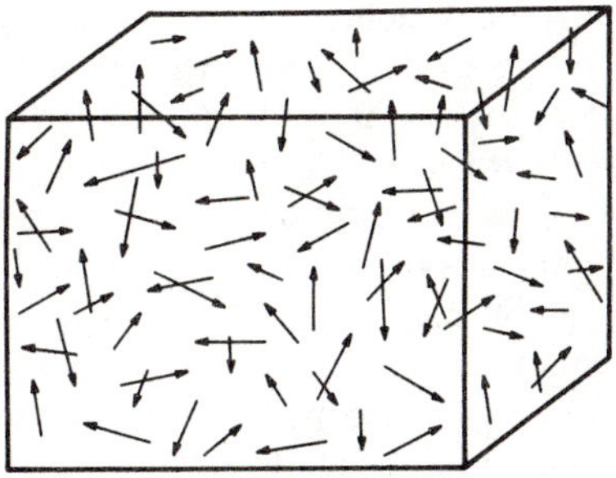

Figure 15

$$\boxed{\begin{aligned} E_{\text{pot}} &= -\sum_i \mathbf{p}_i\cdot\boldsymbol{\mathcal{E}} \qquad &\text{(dielectric)} \\ E_{\text{pot}} &= -\sum_i \boldsymbol{\mu}_{mi}\cdot\mathbf{B} \qquad &\text{(paramagnetic)} \end{aligned}} \tag{39}$$

That is, the total potential energy of these simple systems equals the sum of the potential energies of each molecule — and we ignore the interaction energy between the moments of molecules. A good example of systems in which this interaction energy is very important (and which we shall *not* study) are the ferromagnetic materials. In these substances the magnetic moments interact strongly and tend to make all the $\boldsymbol{\mu}_m$ line up in parallel directions — so the material has a permanent magnetic field associated with it. No attempt will be made here to describe such systems, for as yet there is no complete theory of the thermal behavior of ferromagnetic substances.

5. THE QUANTIZATION OF ENERGY

Prior to the twentieth century Newton's laws of mechanics ("classical" mechanics) were considered to be one of the best experimentally verified theories of physics. Because of the success of this theory in describing planetary motions and numerous other applications, it was naturally assumed to be applicable to all mechanical phenomena. However, near the turn of the century it became apparent that certain "troublesome" experimental observations could not be explained in terms of the existing theories. At first these problems were restricted to the interaction between radiation and matter, which led Max Planck to his revolutionary idea of the quantization of energy, and to certain discrepancies between the theoretical and observed heat capacities of polyatomic gases. In a lecture delivered in 1900 Lord Kelvin described this latter deficiency of the existing statistical theories as a "black cloud" that hung over physics. Some people felt that this black cloud arose from some erroneous fundamental assumptions in the statistical theory of gases, which had been developed by Ludwig Boltzmann. Had they been right, you would have been spared reading this book! However, it turned out that the statistical concepts were not at fault, but rather the underlying concepts of classical mechanics. The theory that Max Planck developed to explain the observed radiation from systems in thermal equilibrium forced him finally to conclude (however reluctantly) that energy could only be exchanged in certain finite ("quantized") amounts. The further development of this idea over the first thirty years of this century culminated in the theory of quantum mechanics, which can be formulated in terms of the Schrödinger equation. Not only did quantum mechanics dissipate Kelvin's black cloud, but it explained many other observed phenomena connected with atomic physics. Obviously the present book is not the place to examine all the implications of quantum mechanics, nor shall we succumb to the common "quickie" derivation of the pertinent results. Instead we shall simply record some of the results of quantum mechanics, indicate the magnitudes of the energy "jumps," and compare these with the classical results. The derivation of these quantum-mechanical results can be found in any good elementary book in quantum mechanics (see the references at the end of this chapter).

Before proceeding to these details, we should note that the distinction between the classical theory and the quantum theory only becomes important when one tries to exchange a small amount of energy with the system. Classically it is always possible to exchange a small amount of energy with a system, whereas the quantum theory tells us we must have a certain minimum amount of energy — namely, the energy difference between two of the quantum states of the system. An analogous situation would be a bank that allows deposits and withdrawals only in units of $1,000. For a millionaire this bank would present no problem, but it is obviously not a bank for a professor. He had better look for a "classical" bank, which allows transactions of any meager amount. Similarly, if energy is added to a system (by heating it up, and thereby raising its temperature), it becomes a "millionaire" (in energy) and can easily afford the quantum jumps in energy, whereas at lower temperatures these transitions may represent a considerable obstacle. Hence we can anticipate that at lower temperatures, where energy is relatively scarce, the quantum restrictions will become more important. Now let us turn to the quantum restrictions of the translational, rotational, and vibrational forms of the energy.

Consider first the case of a *free particle* (a particle that has no force acting on it except at the walls of a container). Let the container have sides of length L_x, L_y, and L_z. Then, according to Schrödinger's equation, the possible values of the translational energy of this particle are

$$\boxed{\epsilon_{\text{trans}} = \frac{h^2}{8m}\left(\frac{n_x^2}{L_x^2} + \frac{n_y^2}{L_y^2} + \frac{n_z^2}{L_z^2}\right)} \tag{40}$$

where m is the mass of the particle and h is Planck's constant ($h = 6.626 \times 10^{-27}$ erg-sec). The quantities (n_x, n_y, n_z) are all positive integers, which may be the same or different. For each different set of values (n_x, n_y, n_z) we say that the particle is in a different quantum state, and these numbers are called the **translational quantum numbers.** This quantum expression for the translational (kinetic) energy is very similar to the classical expression

$$\epsilon_{\text{trans}} = \frac{1}{2}m(v_x^2 + v_y^2 + v_z^2) \equiv \frac{1}{2m}[(mv_x)^2 + (mv_y)^2 + (mv_z)^2]$$

except that the allowed values of the components of the momentum $m\mathbf{v}$ are now discrete. It can be seen that if the container is large (so that L_x, L_y, and L_z are large), then the values of ϵ_{trans} will be very close together and therefore the energy can be treated as continuous. For example, consider a hydrogen atom ($m = 1.66 \times 10^{-24}$ gm) in a cubical container of 1 cm^3 ($L_x = L_y = L_z = 1$). Assume that this atom is in a translational state described by $n_x = 5 \times 10^7$, $n_y = 0$, $n_z = 0$. Then its momentum in the x direction is $n_xh/2L_x \simeq 1.66 \times 10^{-19}$ gm-cm/sec (which corresponds to a velocity of 10^5 cm/sec). Moreover, it has an energy of $\epsilon_{\text{trans}} = (1.66 \times 10^{-19})^2/2 \times 1.66 \times 10^{-24} = 8.3 \times 10^{-14}$ ergs. If n_x is changed by unity, the change in the momentum is only 3.31×10^{-27} gm-cm/sec (which corresponds to a change in the velocity of only 0.002 cm/sec). The change in the energy is $\Delta\epsilon_{\text{trans}} \simeq (h^2/8m)(2n/L_x^2) = (2/n)\epsilon_{\text{trans}} = 4 \times 10^{-8}\epsilon_{\text{trans}}$ (i.e., one twenty-five millionth of the energy)! We see that the translational energy difference between different quantum states is characteristically less than 10^{-20} ergs, and hence the *translational energy can be treated as continuous.*

Next, let us consider the rotational energy of molecules. Here the quantum effects are somewhat more important. According to the quantum theory the allowed values of the rotational energy of a diatomic molecule, such as HCl, are

$$\boxed{\epsilon_{rot} = \left(\frac{h^2}{8\pi^2 I}\right) J(J+1) \qquad (J = 0, 1, 2, \ldots)} \tag{41}$$

where I is the moment of inertia of the molecule. The quantity J is referred to as the **rotational quantum number.** This can also be compared with the classical expression for the rotational energy

$$\epsilon_{rot} = \frac{1}{2} I\omega^2 = \frac{1}{2I}(I\omega)^2$$

(here $\omega^2 = \omega_1^2 + \omega_2^2$). Thus, the quantum theory requires that the total angular momentum $I\omega$ can only have the rather curious values $(h/2\pi)\cdot\sqrt{J(J+1)}$. Consider, for example, the case of HCl, for which $I = 2.65 \times 10^{-40}$ gm/cm² (see Table 3). The difference in ϵ_{rot} between the state $J = 0$ and $J = 1$ is $\Delta\epsilon_{rot} = 2h^2/8\pi^2 I = (6.63 \times 10^{-27})^2/39.4 \times 2.65 \times 10^{-40} = 4.2 \times 10^{-15}$ ergs. (The corresponding classical change in the angular velocity ω would be from $\omega = 0$ to $\omega \simeq 6 \times 10^{12}$ radians/sec.) We see that this jump in energy is about 100,000 times larger than in the translational case (10^{-15} ergs, compared with 10^{-20} ergs). Thus the quantum effects are much more important in restricting the rotational motion than in restricting the translational motion.

The rotational quantum states of molecules are quite involved, but many of these details need not concern us. One point, however, should be noted for future purposes. The rotational quantum state of a molecule is not specified simply by the value of J, for there are a number of states that all have the same value of J. For any fixed value of J the angular momentum along *any fixed direction* can, according to quantum mechanics, have the values $J\hbar$, $(J-1)\hbar$, $(J-2)\hbar, \ldots, -J\hbar$, where $\hbar = h/2\pi$. This means that there are $(2J+1)$ *possible orientations* of the angular momentum, when the total angular momentum is $\sqrt{J(J+1)}\,\hbar$. This is illustrated in Figure 16 (where the five orientations for $J = 2$ are shown). The rotational energy of each of these $(2J+1)$ states is the same and is given by Equation (41).

Of even greater importance than the quantization of ϵ_{rot} is the restriction that quantum mechanics places on the vibrational motion of atoms. According to Schrödinger's equation the possible energies of a **harmonic oscillator** are

$$\boxed{\epsilon_{vib} = (n + \tfrac{1}{2})h\nu \qquad (n = 0, 1, 2, \ldots)} \tag{42}$$

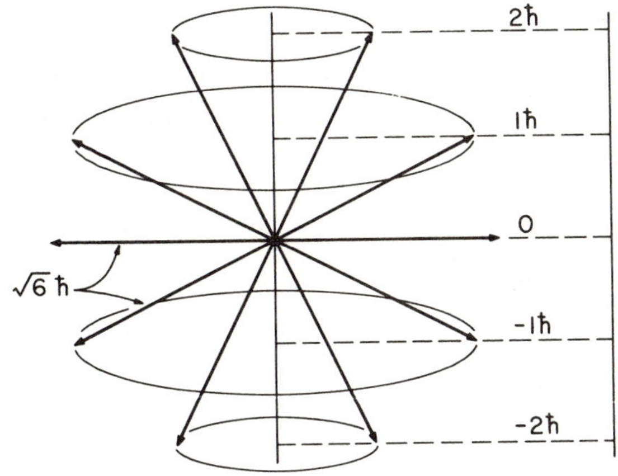

Figure 16

where ν is the "characteristic frequency" of the oscillator (a fixed number which depends on the physical situation). According to this expression the lowest value of the energy of a harmonic oscillator is $\frac{1}{2}h\nu$ (for $n = 0$), which is known as the *zero-point energy*. As we have noted several times, a constant term (such as $\frac{1}{2}h\nu$) can always be neglected in the energy if we simply redefine what we mean by "zero energy" (remember that only the *difference* in energies can be measured). However, in writing Equation (42) the state of zero energy has been selected to be that state in which the oscillator is at rest (classically this corresponds to a particle sitting at its equilibrium position where $\partial\phi/\partial x = 0$). Thus the significance of the zero-point energy is its indication that even when the oscillator has its lowest values of the energy it cannot be at rest — and the term $\frac{1}{2}h\nu$ is retained in order to remind us of this fact. The zero-point energy has *no* effect on the difference in vibrational energy between two states $\Delta\epsilon_{vib}$.

If the vibrational motion of diatomic molecules is treated in the harmonic approximation, then the appropriate values of the characteristic frequency can be determined experimentally. The value of ν (in sec^{-1}) for several diatomic molecules is given in Table 4. Classically the value of ν is

Table 4

HCl,	$\nu = 8.65 \times 10^{13}$	Cl_2,	$\nu = 1.68 \times 10^{13}$
CO,	$\nu = 6.45 \times 10^{13}$	O_2,	$\nu = 4.66 \times 10^{13}$

given by $\nu = (1/2\pi)(\kappa/\mu)^{1/2}$, where κ is the harmonic force constant and μ is the reduced mass of the molecule. As an example of the energy difference between two vibrational states, consider HCl. In this case the difference in energy is

$$\Delta\epsilon_{vib} = 8.65 \times 10^{13} \times 6.63 \times 10^{-27} \simeq 5.7 \times 10^{-13} \text{ erg}$$

which is one hundred times larger than the value we found for $\Delta\epsilon_{rot}$ in HCl. Thus the vibrational motion of a molecule is considerably more restricted than the rotational motion. The relative separation in the energy levels for ϵ_{trans}, ϵ_{rot}, and ϵ_{vib} in diatomic molecules is illustrated (schematically) in Figure 17. The last column in Figure 17 illustrates the difference in energies

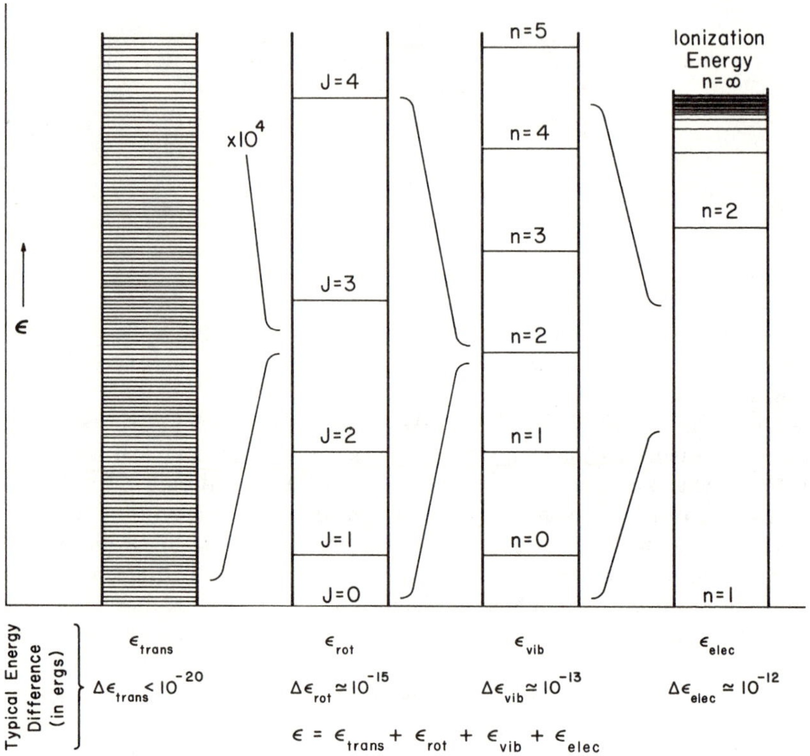

Figure 17

between the electronic levels in a hydrogen atom. The quantum-mechanical expression for these energy levels is

$$\epsilon_{elec} = -\frac{21.76 \times 10^{-12}}{n^2} \text{ (ergs)} = -\frac{13.6}{n^2} \text{ (ev)}$$

where $n = 1, 2, 3, \ldots$. (The kinetic energy acquired by an electron that moves across a potential drop of 1 volt $\equiv$ 1 electron volt (1 ev) $= 1.6 \times 10^{-12}$ erg.) We see that the energy difference, $\Delta\epsilon_{elec}$, is roughly 10^{-12} to 10^{-13} erg, which is quite large. Therefore ϵ_{elec} tends to be constant (and hence can be ignored) unless there is a large amount of energy in the system.

In the last section we saw that vibrational energy is also a major portion of the energy of a solid. It will be recalled that a simple model of a solid is to view it as a collection of $3N$ independent harmonic oscillators (if there are N atoms). If $\epsilon_{\text{vib}}(i) = (n_i + \frac{1}{2})h\nu_i$ is the vibrational energy of the ith oscillator, then the total vibrational energy of the solid is

$$E_{\text{vib}}(n_1, n_2, \ldots, n_{3N}) = \sum_{i=1}^{3N} (n_i + \tfrac{1}{2})h\nu_i \tag{43}$$

The quantum state of the system is specified when the values of the $3N$ **vibrational quantum numbers,** $n_i = 0, 1, 2, \ldots$, are given. The only question remaining is what frequencies ν_i we should assign to these harmonic oscillators. There are two standard methods for assigning values to the frequencies, which lead to slightly different quantum-mechanical models of solids.

The first method, and certainly the simplest possibility, is to assume that all the ν_i's are equal to a common value ν. The value of ν depends on the type of solid in some fashion. This model reduces the expression (43) for ϵ_{vib} to the simpler form

$$\boxed{E_{\text{vib}}(n_1, \ldots, n_{3N}) = \sum_{i=1}^{3N} (n_i + \tfrac{1}{2})h\nu \qquad \text{(Einstein model)}} \tag{44}$$

This model, which was proposed by Einstein in 1906, is commonly referred to as the Einstein model of a solid. It is more accurate for monatomic solids than for diatomic solids (such as NaCl), because it attempts to describe the solid in terms of a single frequency.

The second model, which was proposed by Debye in 1912, accounts for the motion of the atoms in a solid in a more realistic fashion. The essential point is that the atoms of a solid do not really vibrate independently, because the motion of one atom affects the motion of its neighboring atoms. In fact the motion of all the atoms in a solid can be more accurately pictured as a collection of $3N$ standing waves — just as the motion of the atoms in a string can be described by standing waves. Since the amplitude of each wave varies periodically in time, just like a harmonic oscillator, we can again picture the solid as a collection of $3N$ harmonic oscillators, except that now their frequencies are determined by the frequencies of the $3N$ standing waves. The frequency ν of a wave is related to its wavelength λ by $C = \lambda\nu$, where C is the velocity of the wave. If the waves have a common velocity, then the shorter wavelengths correspond to higher frequencies (a short violin string produces a high note). What we must now determine is the number of oscillators (waves) that have a frequency in the range $d\nu$ of ν.

Consider a standing wave inside an enclosure with sides of length L_x, L_y, and L_z in the (x, y, z) directions. Such a wave is described by a function of the form

$$A \sin\left(\frac{n_x \pi x}{L_x}\right) \sin\left(\frac{n_y \pi y}{L_y}\right) \sin\left(\frac{n_z \pi z}{L_z}\right) \sin(2\pi\nu t)$$

where (n_x, n_y, n_z) are positive integers (so the wave vanishes at $x = 0, L_x$, and $y = 0, L_y$, and $z = 0, L_z$). The form of the wave along the x axis is illustrated in Figure 18 for several values of n_x. The solid and dashed curves are for two times that differ by one-half the period of the wave, $(1/2\nu)$. The

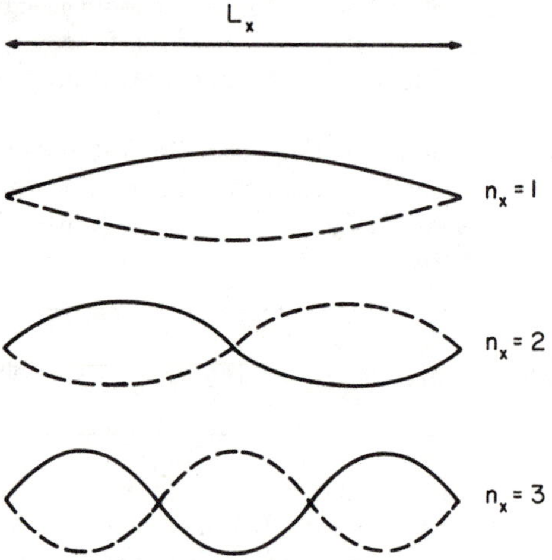

Figure 18

quantities $(q_x, q_y, q_z) = (n_x/2L_x, n_y/2L_y, n_z/2L_z)$ are called the components of the wave number, which has the magnitude $q = (q_x^2 + q_y^2 + q_z^2)^{1/2}$. The velocity of the wave C is related to q and the frequency ν by the condition

$$\nu = qC \qquad \left(\text{that is, } q = \frac{1}{\lambda}\right)$$

From this we see that, since $d\nu = C\,dq$ (C is the same for all waves), the number of these waves with a frequency in the range $d\nu$ of ν is related to the number of waves with a wave number in the range dq of q. To determine the number of waves with a wave number in the range dq of q, we consider a space with coordinates (q_x, q_y, q_z), illustrated in Figure 19. The allowed values of $(q_x, q_y, q_z) = (n_x/2L_x, n_y/2L_y, n_z/2L_z)$ can be represented by points in this space — and these points form a simple cubic-type lattice. Each of these "cubes" has a "volume" of $(1/2L_x)(1/2L_y)(1/2L_z) = 1/8L_xL_yL_z =$

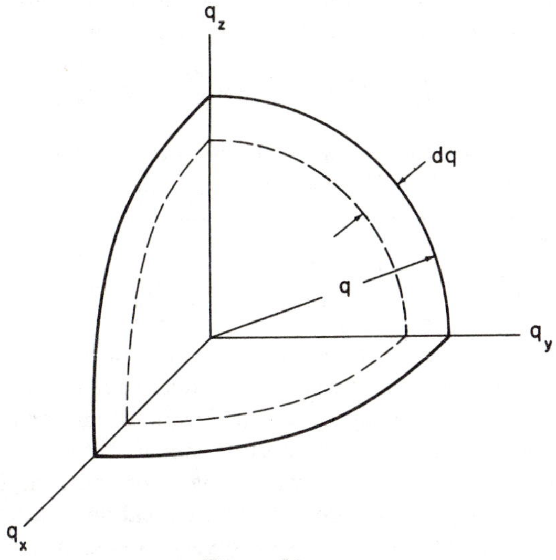

Figure 19

$1/8V$, where V is the volume of the enclosure. Since there is one cube for each point in this space, the volume per point (or wave) is $1/8V$. Now the volume in the spherical shell of thickness dq and radius q is simply $(\frac{1}{8})4\pi q^2\,dq$ [the factor $\frac{1}{8}$ is given because this is one-eighth of a complete spherical shell, since all (q_x, q_y, q_z) are positive]. Dividing this by $(1/8V)$, we find that the number of waves in this shell is $4\pi V q^2\,dq$. Finally, using the fact that $q = \nu C$, we find that

the number of waves with a frequency in the range $d\nu$ of ν is $(4\pi V/C^3)\nu^2\,d\nu$ (45)

where V is the volume of the enclosure and C is their common velocity. This result is rigorously true only if the possible frequencies are continuous, but it is a good approximation if the discrete frequencies are very close together.

There is one final minor complication, owing to the fact that a solid can sustain both longitudinal waves (sound waves) and transverse waves (shear waves). There are two perpendicular transverse waves with some velocity C_t, whereas there is one longitudinal wave with velocity C_l. Thus the number of modes with a frequency in the range $d\nu$ of ν is

$$dN(\nu) = 4\pi\nu\left(\frac{1}{C_l^3} + \frac{2}{C_t^3}\right)\nu^2\,d\nu \equiv \left(\frac{12\pi\nu}{C^3}\right)\nu^2\,d\nu$$

where we have simply set $3/C^3 \equiv C_l^{-3} + 2C_t^{-3}$. Because the solid contains N atoms, and hence has $3N$ degrees of freedom, it can be described in terms of $3N$ waves. There is therefore some maximum frequency ν_m for all these waves, which is determined by the requirement that

$$3N = \int_0^{\nu_m} dN(\nu) = \frac{12\pi V}{C^3}\int_0^{\nu_m} \nu^2\,d\nu = \frac{4\pi V}{C^3}\nu_m^3$$

Thus the *Debye model treats the possible frequencies as continuously distributed between* $0 \le \nu \le \nu_m = (3NC^3/4\pi V)^{1/3}$, *so that all sums of the form*

$$\boxed{\sum_{i=1}^{3N} F(\nu_i) \longrightarrow \int_0^{\nu_m} F(\nu)\,dN(\nu) = \frac{12\pi V}{C^3}\int_0^{\nu_m} F(\nu)\,\nu^2\,d\nu} \qquad (46)$$

where $F(\nu_i)$ is any function of the frequencies. The energy of each oscillator (wave) is, of course, still given by the quantum expression $(n + \frac{1}{2})\,h\nu$.

Finally, let us consider the results of the quantum-mechanical treatment of simple dielectric and paramagnetic systems. This subject is quite involved, but fortunately we can ignore many of the complications in this introductory discussion. Whereas in the classical description the simple dielectric and paramagnetic systems have very analogous expressions for the energy, namely,

$$E = -\sum_i \mathbf{p}_i\cdot\boldsymbol{\mathcal{E}}; \qquad E = -\sum_i \boldsymbol{\mu}_{mi}\cdot\mathbf{B}$$

the quantum-mechanical expressions for the energy are not as similar for the two systems. The essential difference between the two cases is that the magnetic moment of a molecule $\boldsymbol{\mu}_m$ is related to the rotational motion of electrical changes, whereas the dipole moment $\mathbf{p}$ does not depend on rotational motion but only on the separation of charges. As we have already seen, quantum mechanics places restrictions on the rotational energy ϵ_{rot} (or, what is the same thing, on the angular momentum), and these restrictions are reflected in the fact that the magnetic moment can acquire only certain values. To be specific, consider the three substances listed in Table 5.

Table 5

Substance	*Active Ion*	*J*
Gadolinium sulfate, $Gd_2(SO_4)_3\cdot 8H_2O$	Gd^{+++}	$\frac{7}{2}$
Ammonium iron alum, $Fe_2(SO_4)_3(NH_4)_2SO_4\cdot 24H_2O$	Fe^{+++}	$\frac{5}{2}$
Potassium chrome alum, $Cr_2(SO_4)_3K_2SO_4\cdot 24H_2O$	Cr^{+++}	$\frac{3}{2}$

In these rather elaborate molecules it is only the ions Gd^{+++}, Fe^{+++}, and Cr^{+++} that produce the paramagnetic effects. It will be recalled that the total angular momentum of a molecule is given by $(h/2\pi)\sqrt{J(J+1)}$, and the value of J for these various ions is given in the last column of Table 5.

We also discussed the fact that the angular momentum along *any fixed direction* could only have the values

$$(-J, -J + 1, -J + 2, \ldots, J - 1, J)\hbar$$

Now the energy of these molecules in a magnetic field is $\boldsymbol{\mu}_m \cdot \mathbf{B}$, and therefore we are interested in the value of $\boldsymbol{\mu}_m$ *along the direction of* $\mathbf{B}$. But since the value of $\boldsymbol{\mu}_m$ along $\mathbf{B}$ is proportional to the angular momentum along $\mathbf{B}$, it may also only have certain discrete values, which *for the above ions* is given by

$$\boxed{\begin{aligned} \boldsymbol{\mu}_m \cdot \mathbf{B} &= -2\mu_B M B \qquad (-J \leq M \leq J) \\ 2\mu_B &= \frac{eh}{2\pi m} = 1.854 \times 10^{-23} \text{ amp-m}^2 \end{aligned}} \tag{47}$$

(e.g., for gadolinium sulfate $M = -\frac{7}{2}, -\frac{5}{2}, -\frac{3}{2}, -\frac{1}{2}, \frac{1}{2}, \frac{3}{2}, \frac{5}{2}, \frac{7}{2}$). Thus the total energy of these paramagnetic systems is $E = \sum_i \boldsymbol{\mu}_{mi} \cdot \mathbf{B}$, but now the possible values of $\boldsymbol{\mu}_m \cdot \mathbf{B}$ for each molecule are limited to the values given by Equation (47) (μ_B is known as the Bohr magneton).

To illustrate the energy difference between two quantum states, consider the case where $B = 1$ weber/m^2 ($= 10^4$ gauss), which is a strong magnetic field. Then the difference in energy for two quantum states in which M differs by unity is $1.854 \times 10^{-23} \times 1 \times 1 = 1.854 \times 10^{-23}$ joules $= 1.854 \times 10^{-16}$ ergs. This may be compared with the energy differences shown in Figure 17.

The behavior of the simple dielectric system is easier to treat than the paramagnetic system, because quantum effects can generally be ignored. The primary reason for this is, as we have already noted, that the electric dipole does not depend on rotational motion of charges and hence the orientation of $\mathbf{p}$ is not quantized. Moreover it turns out that the rotational motion of the dipole (which is quantized) plays no part in the physical properties, which we shall consider in a later chapter. Therefore we are in the happy position of being able to ignore quantum effects (for reasons that are actually quite subtle) and thereby treat simple dielectrics by classical mechanics. This means that the energy of a dipole in an electric field can be taken to be $-\mathbf{p} \cdot \boldsymbol{\mathcal{E}}$, and the orientation of $\mathbf{p}$ is not quantized.

REFERENCES

G. M. Barrow, *The Structure of Molecules* (New York: W. A. Benjamin, Inc., 1964). An elementary description of the dynamics and quantum conditions for a number of molecules.

T. L. Hill, *Matter and Equilibrium* (New York: W. A. Benjamin, Inc., 1966). A very good introductory discussion of the various equilibrium states of matter (solid, liquid, and gaseous); the interaction energy between molecules, and their relationship to thermodynamic variables and processes.

F. K. Richtmyer, E. H. Kennard, and T. Lauritsen, *Introduction to Modern Physics*, 5th ed. (New York: McGraw-Hill Book Company, 1955).

R. L. Sproull, *Modern Physics*, 2nd ed. (New York: John Wiley & Sons, Inc., 1963).

ESSENTIAL POINTS

1. The total energy of a system of N particles is defined in classical mechanics by the expression

$$E = \sum_{i=1}^{N} \frac{1}{2} m\mathbf{v}_i^2 + \sum_{i=1}^{N} \phi(\mathbf{r}_i) + \frac{1}{2} \sum_{i \neq j} \Phi(r_{ij})$$
$$= E_{\text{kin}} + E_{\text{pot}} + E_{\text{int}}$$

   where E_{kin} is the total kinetic (or translational energy), E_{pot} is the total potential energy due to external forces and E_{int} is the interaction energy between the particles. The total energy does not change in time (it is "conserved").

2. For two *macroscopic* systems A and B the total energy of the composite system E_{A+B} is essentially equal to the sum E_A and E_B. Although E_{A+B} is a constant, E_A and E_B will change in time because of the (weak) interaction between the systems.

3. A perfect gas is characterized by the fact that the interaction energy between the *molecules* can be neglected.

4. The essential feature of a solid is that the atoms vibrate about fixed spatial positions, and therefore the system behaves like a collection of $3N$ harmonic oscillators.

5. Whereas in classical mechanics the energy can have any value, in quantum mechanics the energy is restricted to certain discrete values. The distinction between these two theories becomes important if the energy in the system is small (i.e., the temperature is low) compared with the difference in the allowed values of the energy. Typically, for a molecule, the energy differences between different translational, rotational, vibrational, and electronic states are $\Delta\epsilon_{\text{trans}} \lesssim 10^{-20}$ ergs, $\Delta\epsilon_{\text{rot}} \simeq 10^{-15}$ ergs, $\Delta\epsilon_{\text{vib}} \simeq 10^{-13}$ ergs, $\Delta\epsilon_{\text{elec}} \simeq 10^{-12}$ to 10^{-11} ergs.

PROBLEMS

1. An atom of mass 2×10^{-26} kgm has a kinetic energy of 4×10^{-21} joules when it is in the vicinity of the ground. If it moves to a height of 300 m in the gravitational field, what is its kinetic energy at that point?

2. An electron ($m = 9.1 \times 10^{-31}$ kgm, $e = 1.6 \times 10^{-19}$ coulomb) is acted on by a uniform electric field $\mathcal{E}$ in the positive x direction. Thus, the force on the electron is in the negative x direction. The electric field is produced by two charged plates located at $x = 0$, $x = D$.

   (a) Obtain an expression for the potential energy of the electron. Set $\phi(x = 0) = 0$.

   (b) If an electron has zero velocity when it is at $x = D$, determine its kinetic energy when it reaches $x = 0$ if $\mathcal{E} = 100$ volts/cm, $D = 5$ cm.

   (c) If a particle having 2,000 times the mass of the electron were used in place of the electron in (b), how would its final kinetic energy compare with that of the electron? How would its final velocity compare with the electron's?

3. (a) Show that, in general, if two charges $(+q, -q)$ change their separation distance by an infinitesimal amount $d\mathbf{r}$, then the work done on them by an electric field $\boldsymbol{\mathcal{E}}$ is $\boldsymbol{\mathcal{E}} \cdot d\mathbf{p}$. To show this, consider the work done on each charge separately when they move a distance $d\mathbf{r}_+$ and $d\mathbf{r}_-$, respectively (then relate $d\mathbf{r}$ to these displacements). Use a diagram similar to Figure 3.

   (b) Using the result of part (a), show that if the electric field is increased from zero to $\mathcal{E}$, the work done by the field in forming the *induced* dipole is $\frac{1}{2}\alpha\mathcal{E}^2$ [thus obtaining the second term in Equation (14)].

   (c) Using Equation (8), explain the reason for the minus signs in Equation (14).

4. Determine the polarizability α of two charges $(+q, -q)$ that are bound together by a harmonic potential with a force constant κ.

5. (a) How much work is done on a water molecule if $\mathcal{E}$ is increased from zero to 100 volts/cm and the permanent dipole moment is rotated from $\theta = 0$ to $\theta = \pi$?

   (b) How much work is required to rotate $\boldsymbol{\mu}_m$ ($= 3 \times 10^{-23}$ amp-m^2) of an O_2 molecule from $\theta = \pi$ to $\theta = 0$ if $B = 2$ weber/m^2 (20,000 gauss)?

6. Two physically identical macroscopic systems (same number and type of atoms) have a combined energy of 1,000 joules. At a certain time one system has an energy of 490 joules. Give some numerical values for what the energy of each system *might* be at a later time if: (a) the systems are noninteracting, (b) the systems are interacting. Discuss the interaction energy between the systems.

7. Assume that the atom in Problem 1 has the probability $f(z)\,dz$ of being found in the region dz of z above the ground $(0 \leq z \leq \infty)$.

   (a) What is its average height if $f(z) = Ce^{-\beta mgz}$, where β and C are constants?

(b) Determine C in terms of βmg from the fact that $f(z)$ is normalized to unity (Chapter 1).

(c) Determine its average potential energy.

8. A permanent electric dipole **p** is in an applied electric field $\boldsymbol{\mathcal{E}}$. Assume that we do not know the orientation of **p**, but only the probability $f(\theta)\,d\theta$ that it lies within $d\theta$ of the angle (where $\mathbf{p}\cdot\boldsymbol{\mathcal{E}} = p\mathcal{E}\cos\theta$). Determine its average potential energy if (a) $f(\theta) = \frac{1}{2}\sin\theta$ $(0 \leq \theta \leq \pi)$; (b) $f(\theta) = C\sin\theta\exp(+\beta p\mathcal{E}\cos\theta)$, where $\mathcal{E}$ is the electric field, β is a constant, and C depends only on $(Bp\mathcal{E})$. (c) Using the results of Chapter 1, determine C in terms of $\beta p\mathcal{E}$.

9. (a) Determine the magnitude of the force constant κ for two He atoms, using the harmonic approximation of the Lennard-Jones potential and the data in Table 2.

(b) Determine the frequency $\nu = (1/2\pi)(\kappa/\mu)^{1/2}$ — where μ is the reduced mass — for the vibrational motion of a bound pair of He atoms.

10. Repeat Problem 9 for two O_2 molecules. Explain why the result of part (b) differs from that given in Table 4.

11. Using the *approximate* results of Problem 9 and the quantum-mechanical expression for ϵ_{vib}, determine whether or not two He atoms can form a molecule (He_2). Note that these results are only approximate (and, in fact, lead to the wrong conclusion!), but nevertheless they are instructive. Can you guess why this result might be wrong?

3

Statistical Mechanics

1. *THE BASIC ASSUMPTION OF STATISTICAL MECHANICS*

As we have already noted in the introduction, the object of statistical mechanics is to predict the *macroscopic* properties of systems, using certain assumptions concerning the microscopic composition of the system (types of atoms, forces between atoms, and so on), together with certain *statistical assumptions*. It was argued that we need use only statistical information about the system because the macroscopic properties (pressure, temperature, internal energy, and so on) are due to the *average* behavior of all the atoms of the system. To illustrate this point, we shall now consider how "pressure" results from the microscopic interaction between the atoms of the system and a wall (or pressure gauge).

In a box containing a gas, the atoms are moving in all directions. When they strike the wall of the box, their velocity is changed so that they move back toward the interior of the box. In order to alter the atom's velocity, the wall must, of course, exert a force on the atom. An equal and opposite force is then exerted on the wall (action-reaction). Now the number of atoms that strike one square centimeter of the surface in one second is roughly a million billion billion, i.e., 10^{24} (for air at STP)! Obviously, a pressure gauge cannot differentiate between such frequent impacts. Therefore, it does not record the force that results from the reflection of each individual atom, but only the average force due to a vast number of impacts. It is this average force, divided by the area, which we call the "pressure." This example illustrates the general fact that *all macroscopic properties of a system are averages of some microscopic behavior* within the system. Which macroscopic properties should be identified with which average properties is one of the questions answered by statistical mechanics. The example above shows that this identification often is not very difficult.

Before any averages can be computed, we must first obtain the probability of certain "events." Once these probabilities are obtained, it is a simple matter (in principle at least) to compute the desired averages. However, in order to obtain the probability of these "physical events," we must first settle the question:

"What are the physical events?"

To begin with, let us consider a system with a definite number of particles, say N. When the number of particles in the system is constant, it is called a **closed system.** (The study of open systems will be deferred to Section 6.) Now, generally speaking, a physical event is simply the *detailed microscopic description of the physical system* at some instant of time. From the point of view of classical mechanics, the most detailed information we can obtain about a system is the position and velocity of each of the N particles:

$$r_1, r_2, r_3, \ldots, r_N, \qquad v_1, v_2, v_3, \ldots, v_N \tag{1}$$

The fact that it is humanly impossible to actually determine the value of all these variables need not concern us at present. We are only interested here in determining what (ideal) experiment we might perform. The appropriate experiment in the case of quantum mechanics will be discussed shortly.

To simplify the discussion, consider first the case of one particle ($N = 1$). An ideal experiment would then yield the values of the variables

$$x, y, z, v_x, v_y, v_z \tag{2a}$$

Since these values can vary continuously, it is necessary to consider the "infinitesimal" ranges†

$$dx, dy, dz, dv_x, dv_y, dv_z \tag{2b}$$

†In practice these values are determined by the accuracy of the measuring equipment. Classical mechanics assumes that they may be made arbitrarily small. Quantum mechanics, however, requires that this accuracy cannot exceed the value given by the Heisenberg uncertainty principle, e.g., $dx\, dv_x > h/m$. Thus, for an atom, one could have $dx \simeq 0.1$ cm and $dv_x = 1$ cm/sec (and similarly for y and z). For the present we need not be concerned with this point, so that the ranges (2b) can be arbitrarily small.

in order to discuss the probability of these events. Thus the events of interest are the values (2a) within an accuracy (2b). Following the usual procedure in the case of continuous variables, we introduce a distribution function $f(x, y, z, v_x, v_y, v_z)$ to describe the probability of these events. Then

$$f(x, y, z, v_x, v_y, v_z)\, dx\, dy\, dz\, dv_x\, dv_y\, dv_z \tag{3}$$

is the probability that the particle is located between x and $x + dx$; y and $y + dy$; z and $z + dz$, and has a velocity between v_x and $v_x + dv_x$; v_y and $v_y + dv_y$; v_z and $v_z + dv_z$. This is cumbersome to write in detail each time, so (3) will be written in the form

$$f(\mathbf{r}, \mathbf{v})\, d^3r\, d^3v \tag{3}$$

where $d^3r \equiv dx\, dy\, dz$ and $d^3v \equiv dv_x\, dv_y\, dv_z$, and we shall say that (3) represents the probability that the particle is in the region d^3r of $\mathbf{r}$, with a velocity in the range d^3v of $\mathbf{v}$.

If the system contained two particles ($N = 2$), then the experiment would yield the position and velocity of both particles (simultaneously):

$$\mathbf{r}_1, \mathbf{r}_2, \mathbf{v}_1, \mathbf{v}_2 \quad \text{within the range} \quad d^3r_1, d^3r_2, d^3v_1, d^3v_2$$

The probability of these events could then be represented by

$$f(\mathbf{r}_1, \mathbf{r}_2, \mathbf{v}_1, \mathbf{v}_2)\, d^3r_1\, d^3r_2\, d^3v_1\, d^3v_2$$

where the distribution function now depends on twelve variables. In other words, the probability now depends on where each particle is located, and the velocity of each particle. Obviously, if you already have twelve variables, you might as well "shoot the works" and consider N particles (and $6N$ variables)! Hence, for the original system of N particles, we write

$$f(\mathbf{r}_1, \ldots, \mathbf{r}_N, \mathbf{v}_1, \ldots, \mathbf{v}_N)\, d^3r_1 \cdots d^3r_N\, d^3v_1 \cdots d^3v_N \tag{4}$$

for the probability that particles $1, \ldots, N$ are *simultaneously* in the regions d^3r_1 of $\mathbf{r}_1, \ldots, d^3v_N$ of $\mathbf{v}_N$. This will usually be written in the abbreviated form

$$f(\mathbf{r}_1, \ldots, \mathbf{v}_N)\, d^3r_1 \cdots d^3v_N \tag{4}$$

In statistical mechanics the term **microstate** is frequently used to represent these statistical events, which *describe the microscopic state (or "configuration") of the system.* Thus, in classical mechanics, a microstate is specified by the values of the variables $\mathbf{r}_1, \ldots, \mathbf{v}_N$ within an accuracy $d^3r_1, \ldots, d^3v_N$. The probability of this microstate is then given by (4). Sometimes it is useful to picture the microstate in a space whose coordinates are (x, y, z, v_x, v_y, v_z), which is referred to as a *phase space.* Since it is difficult to draw a space with six coordinates, the idea is illustrated in Figure 1 for a hypothetical one-dimensional system. Each point in this space gives the position and velocity of a particle. These points move around as the position and velocity of the particles change with time — so Figure 1 represents

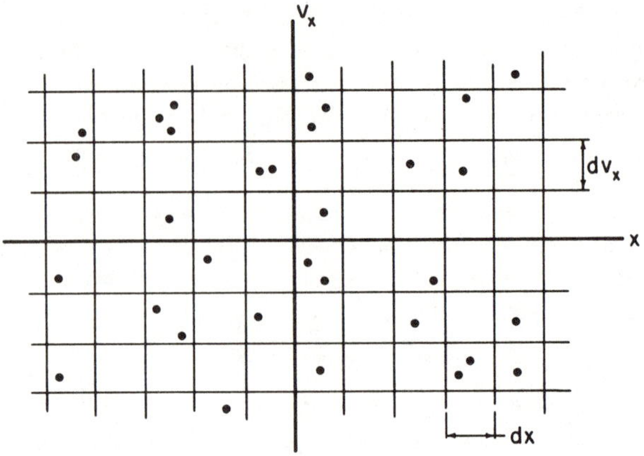

Figure 1

a snapshot at some instant of time. We visualize a gridwork in this space, which marks off regions of size $dx\, dv_x$. Then, from the point of view of classical mechanics, when *any* point moves out of *any* "box," we have a new microstate. It is the probability (4) of these microstates that we want to predict.

The description of a microstate in quantum mechanics is, in certain respects, simpler than the classical description. In quantum mechanics we cannot determine the precise location and velocity $\mathbf{r}_1, \ldots, \mathbf{v}_N$ of the particles (the Heisenberg uncertainty principle requires that $d^3r\, d^3v > h^3/m^3$ for each particle). Instead, the system can be described in terms of certain quantum numbers, which can only have discrete values. What the quantum numbers are, and what they represent in terms of the microscopic state of the system, depends on the particular system (some examples are given in Chapter 2); at this point we do not have to worry about their exact values. A microstate is defined in terms of the values of these quantum numbers. Each different set of values represents a different microstate. We shall designate the energy of a microstate by E_k (the total energy of the system), where k tells us which microstate we are talking about. The value of E_k may be the same for different microstates (different k's), just as the classical energy $E(\mathbf{r}_1, \ldots, \mathbf{v}_N)$ may have the same value for different values of $\mathbf{r}_1, \ldots, \mathbf{v}_N$.

After this rather long-winded discussion about microstates, we turn next to the problem of obtaining their probabilities. It should be fairly evident that the probabilities of the microstates cannot be determined simply by repeated measurements. The number of microstates is so fantastically large that one could never acquire enough data to accurately determine these probabilities. So what can we do? Clearly the only thing to do is to *make some assumption* about these probabilities. We did this frequently in Chapter 1, where we made repeated use of the "reasonable-assumption" method (Equation 4 of Chapter 1) to predict probabilities ("a priori probabilities"). Now we must apply the same argument to the present situation. To do this we must answer the question, "*Which microstates do we feel are equally likely to occur?*" The answer to this question depends on what we know! If we know nothing about the system (not even where it is located!), then all microstates would seem equally likely to occur. However, we generally know that the atoms of the system are located in some container in a definite region of space. In that case, of course, it is not equally likely to find the atoms outside and inside the container, for we know the former situation does not occur. If we know only that the atoms are confined to some region of space, then the most reasonable assumption to make is that all microstates that satisfy this condition are equally probable.

However, we may have one more piece of information about the system besides its spatial location. We may know that the *total energy* of the system has a constant value. The reason that the total energy may be known is that if its value is assigned at one time and the system is isolated (does not exchange energy with other systems), then the energy is known for all times. In other words, the energy of an *isolated system* does not change in time. This is the conservation of energy, discussed in the last chapter. In contrast to this, the microstate of the system is constantly changing with time, so we do not know which microstate occurs at some instant.

Because of this special property of the energy, let us first consider an *isolated system* and assume that we know its total energy and its volume V. Which microstates do we feel have an equal chance of occurring in an experiment? Given only the information above, the most reasonable assumption is that all microstates that satisfy these conditions are equally probable. This "most reasonable" assumption is, in fact, the basic assumption of statistical mechanics.

The Basic Assumption of Statistical Mechanics All microstates of a system that have the same energy are assumed to be equally probable.	(5)

This simple *assumption* is the basis of all of statistical mechanics. Whether or not it is valid is a matter that can only be settled by comparing the predictions of statistical mechanics with actual experiments. To date there is no evidence that this basic assumption is incorrect. A little thought shows that this agreement is indeed remarkable, for our basic assumption is little more than a reflection of our total ignorance about what is going on in the system.

In an isolated system the energy is constant and, according to the assumption (5), all the microstates with this energy are equally probable. A collection (ensemble) of such isolated systems (all identical) is often referred to as a **microcanonical ensemble** — rather a pompous name for a simple concept.† In general we are interested not only in isolated systems, but also in systems that are in thermal contact with heat reservoirs or other systems. In this case the system can exchange energy with the reservoir, so that the energy of the system is no longer constant. An ensemble of such systems (all in thermal equilibrium with reservoirs at the same temperature) is referred to as a **canonical ensemble.** Such an ensemble is beautifully illustrated in Figure 2. The fact that the system can exchange energy with a reservoir leads to a new question, namely: *What is the probability that a*

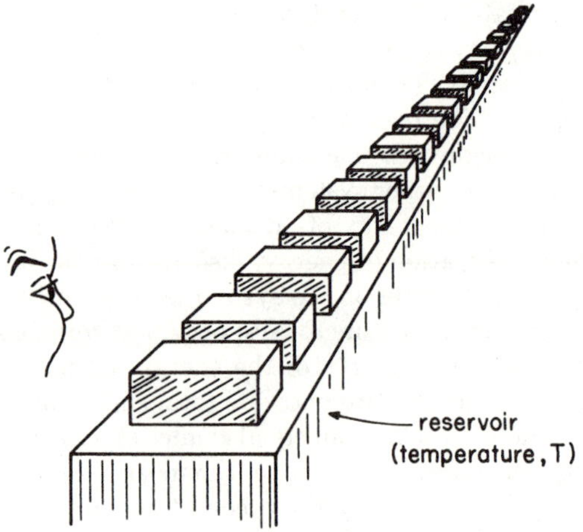

Figure 2

system in a canonical ensemble is in a particular microstate with an energy E? Put another way, if a system can exchange energy with a reservoir, what are the chances of finding it in some particular microstate? We have already decided that all microstates with the same energy are equally probable. This means that the probability of a microstate for a classical system

$$f(\mathbf{r}_1, \ldots, \mathbf{v}_N)\, d^3r \cdots d^3v_N \tag{4}$$

must have the same value for all microstates with the same energy $E(\mathbf{r}_1, \ldots, \mathbf{v}_N)$. This can be so only if the distribution function $f(\mathbf{r}_1, \ldots, \mathbf{v}_N)$

†The adjective "canonical" refers to the fact that this ensemble is the simplest and most significant type. The prefix "micro" refers to the fact that, since the energies of all the systems are equal and constant, this ensemble is a small part of a canonical ensemble.

depends *only* on the energy $E(\mathbf{r}_1, \ldots, \mathbf{v}_N)$. Therefore, the basic assumption of statistical mechanics implies that the probability of a (classical) microstate can be written

$$f[E(\mathbf{r}_1, \ldots, \mathbf{v}_N)]\, d^3r_1 \cdots d^3v_N \tag{6}$$

Similarly, in the case of discrete energies, the probability of a microstate must depend only on the energy E_k of that microstate. We can therefore write

$$P(E_k) = \begin{array}{l}\text{the probability that the system is in a particular} \\ \text{microstate with an energy } E_k\end{array} \tag{7}$$

It should be emphasized that both (6) and (7) refer to the probability of a *particular* microstate — which may be just one of many that have the same energy.

Our previous question can be formulated as:

"How do $f(E)$ and $P(E_k)$ depend on the energy?"

At first blush it might appear reasonable to assume that all values of the energy are equally probable. In this case $f(E)$ and $P(E_k)$ would have to be independent of the energy. A moment's thought shows that this is impossible, for it would be impossible to normalize these functions. For example, we must have

$$\int_V d^3r_1 \cdots \int_{-\infty}^{\infty} d^3v_N\, f[E(\mathbf{r}_1, \ldots, \mathbf{v}_N)] = 1$$

(the sum of the probabilities equals one). However, if $f(E)$ were independent of E, it could be removed from the integrand, and the velocity integrals would be infinite (e.g., $\int_{-\infty}^{\infty} dv_x = v_x|_{-\infty}^{\infty} = \infty$), so this equation could not be satisfied. Therefore, the answer is not this trivial. In particular, we must take into account the fact that the system is in thermal equilibrium with a heat reservoir.

2. DETERMINATION OF $P(E_k)$ AND $f(E)$

To determine $P(E_k)$ [and similarly, $f(E)$] we consider two macroscopic systems, A and B, which are in thermal equilibrium with the same heat reservoir. The probability of finding system A in a particular (micro)state† with energy E_A is, by definition, $P(E_A)$ (that is, it depends only on the energy of the state). Because we are now dealing with more than one system, let us denote this probability by $P_A(E_A)$, so that the probability may depend on the system (e.g., on the types of atoms in the system, and so on). The probability that systems B will be found in a particular state with energy E_B we shall likewise denote by $P_B(E_B)$.

†Henceforth we shall refer to a "microstate" as simply a "state," provided that there is no possibility of confusing it with a thermodynamic state.

Now let us consider the composite system, made up of the system A and system B. We shall call this composite system $A + B$. The probability that this system is in a particular state with energy E_{A+B} is (again) $P_{A+B}(E_{A+B})$. Moreover, since A and B are macroscopic systems, their interaction energy is very small compared with E_A and E_B (see the discussion at the end of Section 2, Chapter 2). Therefore the energy of the composite system is equal to the sum of the energies of systems A and B, or $E_{A+B} = E_A + E_B$, so we have

$$P_{A+B}(E_{A+B}) = P_{A+B}(E_A + E_B)$$

Stated again, this is the probability that the composite system is in a *particular* state with energy $E_A + E_B$. Put another way, this is the probability that system A is in a particular state with energy E_A and *at the same time* system B is in a particular state with energy E_B. This is the same as the *intersection* of the two events "*system A in some particular state with energy* E_A" and "*system B in some particular state with energy* E_B." In terms of the notation of Chapter 1, we might write this as $P[A(E_A)B(E_B)]$. Recall that, if two events are independent, then the probability of their intersection equals the product of the probabilities of the separate events $P(AB) = P(A)P(B)$. We are now going to argue that the events above (in quotation marks) *are* in fact independent. Remember what situation is being considered. Both systems A and system B are in contact with the same thermal reservoir. They are both free to exchange energy with this reservoir. The energy that one exchanges with the reservoir in no way affects the energy the other system may exchange with this reservoir. Consequently, the two systems act independently of one another. Therefore, the state that system A is in is independent of the state that system B is in at that time. This means that the two events (in quotation marks above) are independent, and the probability of their intersection equals the product of their separate probabilities — which are $P_A(E_A)$ and $P_B(E_B)$ respectively. Therefore

$$P_{A+B}(E_A + E_B) = P_A(E_A)P_B(E_B) \tag{8}$$

Using this result, it is now a simple matter to determine how these probabilities depend on the energy. We first note that E_A and E_B are independent variables, so that we can differentiate Equation (8) with respect to either E_A or E_B, holding the other variable constant. Next we note that for any function $F(x + y)$ which depends on x and y only through their sum $(x + y)$, the derivative of $F(x + y)$ with respect to x equals the derivative of $F(x + y)$ with respect to y.† Therefore, if we differentiate Equation (8)

†Because $\partial F(x + y)/\partial x = [dF(x + y)/d(x + y)]\ \partial(x + y)/\partial x = dF/d(x + y)$, and similarly $\partial F(x + y)/\partial y = [dF(x + y)/d(x + y)]\ \partial(x + y)/\partial y = dF/d(x + y)$ — so they are equal.

with respect to either E_A or E_B, the left-hand sides of the equations must be equal, so that we can equate the right-hand sides and obtain

$$P'_A(E_A)P_B(E_B) = P_A(E_A)P'_B(E_B)$$

where the prime represents the derivative. From this we obtain

$$\frac{P'_A(E_A)}{P_A(E_A)} = \frac{P'_B(E_B)}{P_B(E_B)} \tag{9}$$

The left-hand side of Equation (9) depends only on E_A, whereas the right-hand side depends only on E_B. But E_A and E_B can have any values. In that case, the only way that an equation of the form (9) can be satisfied is for both the left-hand and the right-hand sides to be equal to a quantity that is independent of E_A or E_B. Let us call this quantity $-\beta$. Then

$$\frac{dP_A(E_A)}{dE_A} = -\beta P_A(E_A)$$

or

$$P_A(E_A) = C_A e^{-\beta E_A} \tag{10}$$

C_A is a factor that depends on the composition (atomic) of the system A. On the other hand, the quantity β does *not* depend on the composition of the system A. We can see this by remembering that both the left and right sides of (9) equal $-\beta$, so that β must be a quantity that is common to both systems A and B. But the only thing that system A and B have in common is the fact that they are in thermal equilibrium with the same reservoir. Therefore, the quantity β must have something to do with the reservoir. Since the temperature T is the only quantity that distinguishes one reservoir from another, we expect to find that β is related to T. Equation (10) is also valid for system B (replacing the subscript A with B), so we have obtained the desired result — namely, that

$$P(E) = Ce^{-\beta E} \tag{11}$$

is the probability that a system, in thermal equilibrium with a reservoir, will be found in a particular state with energy E. The factor C depends on the composition of the system, whereas the quantity β depends only on the temperature of the reservoir T. In the case where the energy can vary continuously, we have the same result for the distribution function $f(E)$, Equation (6), namely

$$f[E(\mathbf{r}_1, \ldots, \mathbf{v}_N)] = C'e^{-\beta E(\mathbf{r}_1, \ldots, \mathbf{v}_N)} \tag{12}$$

We shall often write this simply as

$$f(E) = C'e^{-\beta E} \tag{12}$$

Equations (11) and (12) are frequently written in a different form, in which the factors C and C' are replaced by expressions that we shall now obtain. Recall that if the probabilities of all events are added together, the result must equal unity. Hence, summing over all microstates (ms),

$$\sum_{\text{ms}} P(E_k) = \sum_k Ce^{-\beta E_k} = C\sum_k e^{-\beta E_k} = 1$$

or, in the case (12),

$$\int_V \cdots \int_{-\infty}^{\infty} f(E)\, d^3r_1 \cdots d^3v_N = C' \int_V \cdots \int_{-\infty}^{\infty} e^{-\beta E}\, d^3r_1 \cdots d^3v_N = 1$$

The term V on the first integral is to indicate that the spatial integrals extend over the volume of the system. The velocity integrals all extend from $-\infty$ to $+\infty$. In the case of discrete energies it is convenient to set $C = Z^{-1}$, so that

$$\boxed{P(E_k) = Z^{-1}e^{-\beta E_k}, \qquad \text{where } Z = \sum_k e^{-\beta E_k}} \tag{13a}$$

Note that Z is a dimensionless quantity. In order to obtain a similar dimensionless quantity in the classical case we set $C' = D_n Z^{-1}$ and write

$$\boxed{f(E) = Z^{-1}D_N e^{-\beta E}, \qquad \text{where } Z = D_N \int_V d^3r_1 \cdots \int_{-\infty}^{\infty} d^3v_N e^{-\beta E}} \tag{13b}$$

Now Z is dimensionless provided that D_N has the dimensions $(\text{length})^{-3N}$ $(\text{velocity})^{-3N}$. The numerical value of D_N, however, is not measurable, for it does not influence the value of $f(E)$ [note that D_N cancels out of the expression for $f(E)$]. We have temporarily introduced it simply to make Z a dimensionless quantity. From the point of view of classical mechanics there is no reason why we should not take the simplest possible value for D_N, namely

$$D_N = 1 \quad (\text{length-velocity})^{-3N} \tag{14}$$

We shall use this convention† for all systems with a fixed value of N (closed systems), so that D_N will not appear explicitly in (13b). Finally, the quantity Z, which is a function of β and V, is called the **partition function** (Planck referred to it as the *Zustandssumme* — state sum; this is where the Z comes from). At this stage of the game Z is nothing more than a normalizing factor which guarantees that the sum of the probabilities (13) equals unity.

†Another convention, based on quantum-mechanical considerations, is frequently used at this point. This convention goes outside of the framework of classical mechanics; therefore we will not use it in (13b). As we will see later, no problems arise from the convention (14), despite frequent statements to the contrary (e.g., see Appendix D).

Now an important point to note is that Equation (13) gives the probability that the system is in a *particular* (micro)state. Thus, even though $e^{-\beta E}$ is largest for $E = 0$, *this does not imply that the system is most likely to be found with zero energy.* The reason is that there are generally many states of the system with the same value of the energy. The probability that the system *has a certain energy* E (*not* that it is in a *particular* state with energy E) is equal to the sum of the probabilities of all the states with the energy E, or simply

$$\begin{pmatrix}\text{probability that the} \\ \text{system has an energy } E\end{pmatrix} = \begin{pmatrix}\text{number of states} \\ \text{with energy } E\end{pmatrix} \times P(E) \qquad (15)$$

The collection of all the states having the same energy is often called an energy level. Thus an **energy level** is simply a compound event, composed of all states with the same energy. If we denote the probability of an energy level by $P_\ell(E)$, then according to our definition of the probability of a compound event,

$$\boxed{P_\ell(E) = \sum_{E_k = E} P(E_k) = \Omega(E)P(E)} \qquad (15)$$

where

$$\Omega(E) = (\text{number of states with energy } E) \qquad (16)$$

$\Omega(E)$ is often referred to as the *degeneracy*, or the weighting factor, of the energy level E. As we shall soon see, $\Omega(E)$ generally increases with increasing values of E (i.e., there are more states with higher energy). Thus from Equations (15) and (13) we get a picture like the one shown in Figure 3. We see that, even though the probability of a particular state decreases as E

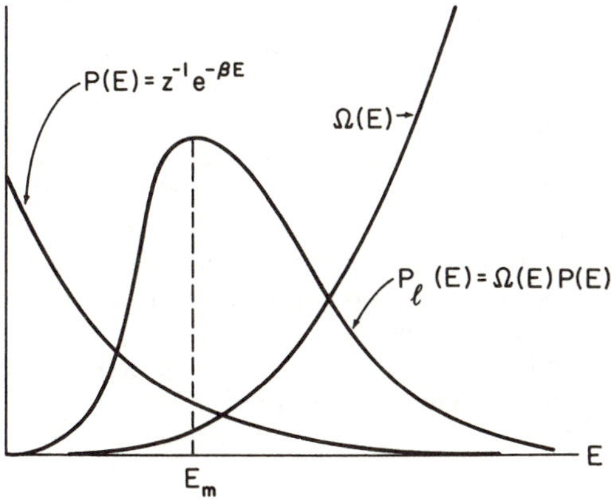

Figure 3

increases, the system is most likely to have an energy $E = E_m > 0$ because $\Omega(E)$ increases with E.

Example

To illustrate the discussion above, consider the hypothetical system of five noninteracting particles, each of which can have only the energies $\epsilon_i = n_i\epsilon_0$ $(n_i = 0, 1, 2, \ldots)$. The total energy of the system is therefore $E = \epsilon_1 + \epsilon_2 + \epsilon_3 + \epsilon_4 + \epsilon_5$. The degeneracy $\Omega(E)$ for this system is determined in the accompanying table. The first column gives the energy of the

Typical Energy Distribution

Energy Levels	*Particles* 1	2	3	4	5	*Number of Arrangements*	$\Omega(E)$
$E = 0$	0	0	0	0	0	1	1
$E = 1\epsilon_0$	ϵ_0	0	0	0	0	5	5
$E = 2\epsilon_0$	$2\epsilon_0$	0	0	0	0	5	15
	ϵ_0	ϵ_0	0	0	0	10	
$E = 3\epsilon_0$	$3\epsilon_0$	0	0	0	0	5	35
	$2\epsilon_0$	ϵ_0	0	0	0	20	
	ϵ_0	ϵ_0	ϵ_0	0	0	10	
$E = 4\epsilon_0$	$4\epsilon_0$	0	0	0	0	5	70
	$3\epsilon_0$	ϵ_0	0	0	0	20	
	$2\epsilon_0$	$2\epsilon_0$	0	0	0	10	
	$2\epsilon_0$	ϵ_0	ϵ_0	0	0	30	
	ϵ_0	ϵ_0	ϵ_0	ϵ_0	0	5	

system. The next five columns give *one* of the possible ways in which the energy can be distributed among the particles. The sixth column gives the total number of distinct arrangements for the indicated particle energies. The resulting degeneracy $\Omega(E)$ satisfies $\Omega(n\,\epsilon_0) = (n + 4)!/n!4!$, the derivation of which need not concern us at present. The important point is that it is a rapidly increasing function of n. Note that $\Omega(E)$ simply results from counting the number of possible states of energy E. It has nothing to do with the probability that these states will be occupied, and nothing to do with β [both of which are contained in $P(E)$].

The partition function for the present system is

$$Z = \sum_{\text{states}} e^{-\beta E} = \sum_{n_1=0}^{\infty} e^{-\beta n_1 \epsilon_0} \sum_{n_2=0}^{\infty} e^{-\beta n_2 \epsilon_0} \times \cdots \times \sum_{n_5=0}^{\infty} e^{-\beta n_5 \epsilon_0}$$

Each of these sums may be put into a simpler form if we use the fact that $(1 - x)^{-1} = \sum_{n=0}^{\infty} x^n$ (where, in the present case, $x = e^{-\beta\epsilon_0}$). Since all sums yield the same value, we have

$$Z = (1 - e^{-\beta\epsilon_0})^{-5}$$

The value of Z is then determined once $\beta\epsilon_0$ is specified. For example, if $\beta\epsilon_0 = \frac{1}{2}$, then $Z \simeq 106$, and using $P(E) = Z^{-1} e^{-\beta E}$ one can obtain $P(E)$. The value of $P_\ell(E) = \Omega(E)\, P(E)$ is then obtained from the values of $\Omega(E)$ in the above table. Figure 3(a), which is the analogue of Figure 3, shows the results of

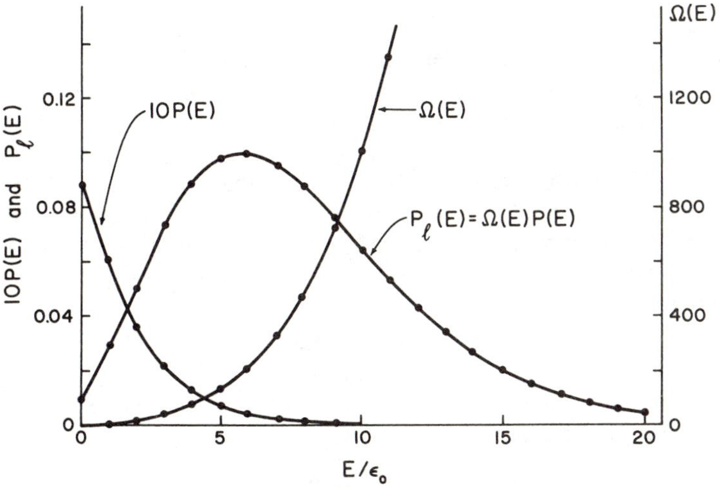

Figure 3(a)

this procedure (when $\beta\epsilon_0 = \frac{1}{2}$). In this case the most probable energy is $E_m = 6\epsilon_0$. Note that the degeneracy is larger than 1,200 when $E = 11\epsilon_0$, but $P(E)$ is so small that $P_\ell(E)$ decreases. A simpler example, involving the same ideas, is given in the problems.

Equation (13), namely $P(E) = Z^{-1}e^{-\beta E}$, is the basic result used for all calculations in statistical mechanics. As we discussed at the beginning of this chapter, the macroscopic properties of a system are related to average microscopic properties, which we can compute only after we have determined the probabilities of the various microstates. Now that we have those probabilities, we can compute any desired averages. But which averages are related to which macroscopic properties? This is the question we take up next.

3. INTERNAL ENERGY

We begin by considering the internal energy. The other thermodynamic variables will be considered in a later section. Obviously, the internal energy U of a system is related in some fashion to "the energy" within a system when it is in contact with a reservoir. However, we have already noted that when the system is in contact with a reservoir, its energy E fluctuates — because it continually exchanges energy with the reservoir. Therefore, it is meaningless to speak of "the energy" of this system. But what then is the internal energy U? When we measure the change in U, we use macroscopic instruments, which are insensitive to all the minute variations that occur at the microscopic level. We noted this insensitivity at the beginning of Section 1 in which we discussed the measurement of pressure. The pressure gauge does not record the force of each atom separately, but only the average force. In the present case, our instruments do not indicate the actual energy exchange between systems at a particular instant of time, but rather the average energy over some time interval (say one second). This is presumably the same as saying that our instrument has averaged the results of a large number of "ideal experiments," which measure the instantaneous energy of the system (the ergodic hypothesis). For these reasons we conclude that the internal energy should be identified with the statistical average energy,

$$U = \bar{E} = \sum_k E_k P(E_k) = Z^{-1} \sum_k E_k e^{-\beta E_k} \tag{17}$$

This result can be expressed in another fashion, which is often very useful. Note that the partition function

$$Z = \sum_k e^{-\beta E_k} \tag{18}$$

depends on the value of β. In other words, it is a function of β. If we differentiate Z with respect to β, we obtain

$$\frac{\partial Z}{\partial \beta} = \sum_k \frac{\partial}{\partial \beta} e^{-\beta E_k} = -\sum_k E_k e^{-\beta E_k}$$

If we divide this by Z and compare it with (17), we see that $Z^{-1}\partial Z/\partial\beta$ equals $(-U)$. But $Z^{-1}\, \partial Z/\partial\beta = \partial(\ln Z)/\partial\beta$, so we find that

$$\boxed{U = -\frac{\partial(\ln Z)}{\partial \beta}} \tag{19}$$

This means that if we know Z, we can determine U simply by taking the logarithm of Z and differentiating it with respect to β. Remember that the parameter β can depend only on the temperature and not on the atomic composition of the system. Therefore, to determine how β is related to the temperature, we need only evaluate Z for some simple system and then compare the known expression for U with the right side of (19). Obviously (or not), the simplest system to consider is a perfect monatomic gas. Let us therefore proceed to consider this system.

4. *THE IDENTIFICATION OF* β†

Real gases behave like perfect gases when their density (pressure) is sufficiently reduced. From a microscopic point of view, we see that when the gas is sufficiently rarefied (by reduction of its density), the atoms in the gas see mostly empty space — only rarely do they collide with one another. This means that the potential energy due to the interatomic forces is essentially zero most of the time. Therefore, the energy of the system is essentially just the sum of the kinetic energies of the atoms,

$$E = \sum_{i=1}^{N} \epsilon_i = \sum_{i=1}^{N} \tfrac{1}{2}m\mathbf{v}_i^2 \tag{20}$$

We conclude that Equation (20) should represent the energy of perfect monatomic gas containing N atoms, all of mass m. If the gas were not monatomic, there would have to be additional terms representing the potential energy between the atoms in each molecule (see Section 4 of Chapter 2). For the time being we shall restrict our considerations to the monatomic gas. If we substitute (20) into (13b) (using $D_N = 1$), then we find that for this system

$$f(E) = Z^{-1}\exp(-\beta E) = Z^{-1}\exp\left(-\beta \sum_{i=1}^{N} \epsilon_i\right), \qquad \epsilon_i = \tfrac{1}{2}m\mathbf{v}_i^2$$

Since the energy is simply a sum of the individual energies ϵ_i, $f(E)$ can be written as a product of terms

$$f(E) = Z^{-1}e^{-\beta\epsilon_1}\,e^{-\beta\epsilon_2}\cdots e^{-\beta\epsilon_N} \tag{21}$$

The partition function simplifies in a similar fashion. Referring again to Equation (13b), we see that

$$Z = \int_V \cdots \int_{-\infty}^{\infty} e^{-\beta\epsilon_1}\,e^{-\beta\epsilon_2}\cdots e^{-\beta\epsilon_N}\, d^3r_1 \cdots d^3v_N \tag{22}$$

†An alternative method for identifying β, which some people prefer, is given in Appendix C. The present method is slightly less abstract, quite general (although it looks special!), and involves some simple but useful analysis. The choice of methods is largely a matter of taste.

But since the energy does not depend on the position of the atoms, the spatial integrals can be performed immediately. Each integral of a variable $\mathbf{r}_i$ yields the volume V of the systems, so

$$Z = V^N \int_{-\infty}^{\infty} d^3v_1\, e^{-\beta\epsilon_1} \int_{-\infty}^{\infty} d^3v_2\, e^{-\beta\epsilon_2} \cdots \int_{-\infty}^{\infty} d^3v_N\, e^{-\beta\epsilon_N}$$

Here we have separated the velocity integrals appearing in Equation (22) and written them in the form of a product of integrals. Note that these integrals are all identical, for they all equal

$$\int_{-\infty}^{\infty} d^3v \exp(-\beta\tfrac{1}{2}m\mathbf{v}^2)$$

therefore

$$\boxed{Z = \mathfrak{z}^N} \tag{23}$$

The quantity $\mathfrak{z}$ is known as the **molecular** (*or particle*) **partition function.** It should not be too difficult to see that the form (23) for the partition function is valid whenever Z can be written in the form (22) — regardless of the form of the particle's energy $\epsilon(\mathbf{r}, \mathbf{v})$ (see problems). In the present case ($\epsilon = \frac{1}{2}m\mathbf{v}^2$) the molecular partition function is

$$\mathfrak{z} = V \int_{-\infty}^{\infty} d^3v \exp(-\beta\tfrac{1}{2}m\mathbf{v}^2) \tag{24}$$

We see that if (23) is substituted into (21), the distribution function for the entire system can be written in the form

$$\begin{aligned} f(E) &= (\mathfrak{z}^{-1} e^{-\beta\epsilon_1})(\mathfrak{z}^{-1} e^{-\beta\epsilon_2}) \cdots (\mathfrak{z}^{-1} e^{-\beta\epsilon_N}) \\ &= f(\epsilon_1)f(\epsilon_2) \cdots f(\epsilon_N) \end{aligned} \tag{25}$$

where the right side contains the distribution functions for the individual atoms. We see that, because the atoms in a perfect gas do not have any interaction energy, they behave statistically as *independent systems* [for the right side of Equation (25) consists of a product of distribution functions]. The probability that any particular atom is located in the region d^3r of $\mathbf{r}$ and has a velocity in the range d^3v of $\mathbf{v}$ is given by

$$\begin{aligned} f(\epsilon)\, d^3r\, d^3v &= \mathfrak{z}^{-1} e^{-\beta\epsilon}\, d^3r\, d^3v \\ &= \mathfrak{z}^{-1} \exp(-\beta\tfrac{1}{2}m\mathbf{v}^2)\, d^3r\, d^3v \end{aligned} \tag{26}$$

where $\mathfrak{z}$ is given by Equation (24), and $\epsilon = \frac{1}{2}m\mathbf{v}^2$. We will discuss this distribution function in detail in the following chapter.

Now let us use these results to determine how β is related to the temperature T of the system. To determine this, we must first determine $Z = \mathfrak{z}^N$ [Equation (23)]. Writing out (24) in terms of the components of the velocity, we have

$$\mathfrak{z} = V \int_{-\infty}^{\infty}\int_{-\infty}^{\infty}\int_{-\infty}^{\infty} dv_x\, dv_y\, dv_z \exp[-\beta\tfrac{1}{2}m(v_x^2 + v_y^2 + v_z^2)]$$

This integral can be written as the product of three integrals in the same way as was done for Equation (22). Since each integral is the same (the variable of integration does not affect the value of the integral), we find that

$$\mathfrak{z} = V\left(\int_{-\infty}^{\infty} dv\, e^{-\beta m v^2/2}\right)^3 \tag{27}$$

The integral that appears in Equation (27) is the same type of Gaussian integral discussed in Chapter 1. It has the value $(2\pi/\beta m)^{1/2}$ (see Appendix A), so

$$\mathfrak{z} = V\left(\frac{2\pi}{\beta m}\right)^{3/2} \tag{28}$$

Substituting this into Equation (23) yields the classical partition function for the perfect monatomic gas:

$$\boxed{Z = V^N \left(\frac{2\pi}{\beta m}\right)^{3N/2}} \tag{29}$$

We shall not use Equation (29) in Equation (19) to determine the internal energy of this system in terms of β. Note first that, in the present case

$$\ln Z = \ln\left[V^N \left(\frac{2\pi}{\beta m}\right)^{3N/2}\right] = N \ln V + \frac{3}{2} N \ln\left(\frac{2\pi}{\beta m}\right)$$

so

$$U = -\frac{\partial}{\partial\beta} \ln Z = -\frac{3}{2} N \frac{\partial}{\partial\beta} \ln\left(\frac{2\pi}{\beta m}\right) = +\frac{3}{2} N \frac{\partial}{\partial\beta} \ln\left(\frac{\beta m}{2\pi}\right) = \frac{3N}{2\beta}$$

Now, according to Equation (19), this must be equal to the internal energy of a perfect monatomic gas. But it is known from experiments that the heat capacity of monatomic gases is $C_V = 3nR/2$ (n is the number of moles and R is the gas constant). For our present purposes it is more useful to write this in the form $C_V = 3Nk/2$, where N is the number of atoms and k is known as the *Boltzmann constant* ($k = 1.380 \times 10^{-16}$ ergs/deg).† Since $C_V = (\partial U/\partial T)_V$, we can set $U = 3NkT/2$ (also see problems). Comparing this with the above expression for the internal energy, we must have

$$\frac{3}{2} NkT = U = \frac{3N}{2\beta}$$

or, in other words,

$$\boxed{\beta = \frac{1}{kT}} \tag{30}$$

†A method for determining the value of this constant is described in Section 4 of the following chapter.

Let us make it clear that the result (30) is entirely general. It has nothing to do with a perfect gas — it holds for all systems. This may seem strange, since in the above derivation we considered only a perfect gas. But it should be recalled that in the derivation of the general expression

$$P(E) = Z^{-1} e^{-\beta E} \tag{13}$$

we showed that β is independent of the composition of the system (the types of atoms, forces, and so on within the system). It follows that we can use *any* system to determine β. The reason we choose a perfect monatomic gas is that the partition function is very simple to evaluate in that case.

5. *THE REMAINING THERMODYNAMIC FUNCTIONS*

We shall now proceed to obtain general expressions for the remaining thermodynamic state functions — the entropy, the free energy, equations of state, and so on. The application of these results to particular systems will be studied in the following chapter.

We have seen that the most convenient way of determining the internal energy is to consider the partition function to be a function of β and to write

$$U = -\frac{\partial \ln Z}{\partial \beta}$$

Now the partition function

$$Z = \int_V \cdots \int_{-\infty}^{\infty} d^3r_1 \cdots d^3v_N \, e^{-\beta E} \qquad \text{or} \qquad Z = \sum_k e^{-\beta E_k}$$

depends not only on β but also on V, the volume of the system. This is apparent in the first expression from the fact that the limits of the spatial integrals depend on the volume of the system. In the second expression for Z (discrete energy states), the dependence on V is not explicit. Nonetheless, Z does depend on V, because the energies E_k are functions of V (how they depend on V need not worry us at present). Moreover, Z depends on N, the number of particles in the system, and on any fields that may be present, such as the magnetic, electric, and gravitational fields. The dependence of Z on these fields is again not explicit but arises from the fact that the energy E depends on their magnitude. For the present, we shall consider only the case in which N is a definite fixed number (closed system) and any fields that are present are not varied. The case when these quantities are variable can easily be treated, but it is not essential for (and in fact complicates) the following discussion.

The partition function $Z = Z(\beta, V)$ now will be treated as a function of both β and V. The change in $\ln Z$, when both β and V are varied, is given by

$$d(\ln Z) = \left(\frac{\partial \ln Z}{\partial \beta}\right) d\beta + \left(\frac{\partial \ln Z}{\partial V}\right) dV = -U\, d\beta + \left(\frac{\partial \ln Z}{\partial V}\right) dV$$

If we write the first term in the form $-U\, d\beta = -d(U\beta) + \beta\, dU$, the last equation becomes

$$d(\ln Z + U\beta) = \beta\, dU + \left(\frac{\partial \ln Z}{\partial V}\right) dV.$$

If we multiply this by kT (remember that $\beta = 1/kT$) and rearrange terms, we find that

$$dU = kT\, d(\ln Z + \beta U) - kT \left(\frac{\partial \ln Z}{\partial V}\right) dV \tag{31}$$

Recall that the combined first and second law of thermodynamics is (for a system with state variables P, V, and T)

$$dU = T\, dS - p\, dV$$

Comparing this with Equation (31), one can only conclude that

$$\boxed{dS = k\, d(\ln Z + \beta U)} \tag{32}$$

$$\boxed{p = kT \left(\frac{\partial \ln Z}{\partial V}\right)} \tag{33}$$

Equation (33) relates the pressure to the other thermodynamic variables, once Z is known. Thus (33) is simply the **equation of state** of the system (any system). Also Equation (32) tells us how the entropy varies, once Z is known. One sees once again why the partition function is such a useful quantity. Once Z is determined, not only can U be determined but also the pressure and entropy of the system.

Although we shall limit most of our applications of these results to the following chapter, it is so simple to obtain the equation of state for a perfect monatomic gas from (33) that it seems nearly criminal to delay this point. Moreover, a trivial application at this stage may make these results more believable, and hence more meaningful. The partition function for a perfect monatomic gas is, according to Equation (29),

$$Z = V^N \left(\frac{2\pi}{m\beta}\right)^{3N/2}$$

and therefore

$$\ln Z = N \ln V + \left(\frac{3N}{2}\right) \ln \left(\frac{2\pi}{m\beta}\right) \tag{34}$$

If we differentiate this with respect to V, only the first term on the right side contributes, and this yields N/V. Consequently, if (34) is substituted into (33), we find that

$$p = \frac{NkT}{V}$$

which indeed is the equation of state for a perfect gas.

We note that Equation (32) only expresses the differential of the entropy in terms of statistical quantities. If we integrate it, we obtain

$$\boxed{S = k \ln Z + k\beta U + \text{constant } (N, \text{field strengths})} \tag{35}$$

For the present the constant in (35) is of no interest. It should be noted, however, that it can depend on N and the magnitude of any fields (all of which have been treated as constant). For the present we shall take this constant to be zero.†

From Equation (35) we can readily determine other thermodynamic functions. Thus the Helmholtz free energy is given by

$$F = U - TS = U - kT \ln Z - U = -kT \ln Z$$

or

$$\boxed{F = -kT \ln Z} \tag{36}$$

The change in the Helmholtz free energy is given by

$$dF \equiv d(U - TS) = -S\,dT - p\,dV$$

which, as can be seen, is most naturally expressed in terms of the change in T and V. If we consider $F(T, V)$ as a function of these variables, we can also write

$$dF = \left(\frac{\partial F}{\partial T}\right)_V dT + \left(\frac{\partial F}{\partial V}\right)_T dV$$

Comparing this with the last expression yields the thermodynamic relationships

$$S = -\left(\frac{\partial F}{\partial T}\right)_V, \qquad p = -\left(\frac{\partial F}{\partial V}\right)_T \tag{37}$$

†Since only the change in the entropy dS can be measured, this constant has no physical significance. However, the entropy is often assumed to satisfy certain additional requirements, in which case either this constant cannot be taken to be zero, or else D_N cannot be given by (14). This point is illustrated in the problems and is discussed in more detail in Appendix D.

It is not difficult to show that if (36) is substituted into (37), one does in fact recover Equations (35) and (33). The fact that the Helmholtz free energy is most naturally expressed in terms of (T, V) is the reason it is so simply related to $Z(T, V)$, Equation (36). The remaining thermodynamic functions are somewhat more involved expressions of Z, which will be left as problems (naturally!).

Finally, it is interesting to note that from (36)

$$Z = e^{-F/kT} \qquad \text{or} \qquad Z^{-1} = e^{+\beta F}$$

so the probabilities of a microstate can also be written in the form

$$Z^{-1}e^{-\beta E} = e^{\beta(F-E)} \tag{38}$$

This form for the probability (and the distribution function) is frequently encountered in the literature on statistical mechanics.

6. ANOTHER LOOK AT ENTROPY

The entropy is a state function which, being the first-born child of the second law of thermodynamics, has a certain aura of mystery. It is not as easily "understood" as, for example, the internal energy, which can be pictured in terms of the mechanical concept of energy. To be sure, the change in the entropy $dS = (dQ)_{\text{rev}}/T$ (in terms of the heat added to a system in a reversible process) is a well-defined quantity which can be measured between any two equilibrium thermodynamic states. This expression, however, does not give us any insight into what S represents in terms of the molecular states of the system. This is to be expected, of course, since thermodynamics is not concerned with the molecular description of systems.

The question now arises about whether statistical mechanics can shed any light on the microscopic significance of the entropy. Certainly the expression that we have found,

$$S = k \ln Z + k\beta U \tag{35}$$

does not look very revealing (we drop the constant in S temporarily, for simplicity). However, we shall now show that Equation (35) can be put into a form that has a fairly simple statistical interpretation.

We begin by considering the unlikely expression

$$-k \sum P(E_k) \ln P(E_k)$$

(or, in the classical case, $-k\int_V \cdots \int_{-\infty} f(E) \ln f(E)\, d^3r_1 \cdots d^3v_N$), and substitute the expression

$$P(E_k) = Z^{-1}e^{-\beta E_k}$$

We then obtain

$$-k\sum P(E_k) \ln (Z^{-1}e^{-\beta E_k}) = +k\sum P(E_k) \ln Z + k\sum E_k\, \beta P(E_k)$$

Since $\Sigma\, P(E_k) = 1$, and $U = \bar{E} = \Sigma\, E_k P(E_k)$, this equals $k \ln Z + k\beta U$, which is the same as Equation (35). We therefore conclude that (aside from an additive constant)

$$\boxed{S = -k \sum P(E_k) \ln P(E_k)} \tag{39}$$

This statistical expression for the entropy, although not particularly useful for calculations, is very useful for the purpose of obtaining a statistical interpretation of the entropy. This interpretation rests on the similarity between Equation (39) and the "uncertainty" H, which was discussed in Section 6 of Chapter 1. In fact, since

$$H = -\sum P(E_k) \ln P(E_k)$$

we have from Equation (39) that

$$\boxed{S = +kH} \tag{40}$$

We can conclude that *the entropy of a thermodynamic* (i.e., equilibrium) *state is proportional to the* **statistical uncertainty** *associated with that thermodynamic state.* It is important to remember that *each* thermodynamic state is characterized in statistical mechanics by the values of the probabilities $P(E_k)$ of *all* the *microstates.* According to the above result, the more uncertain we are about which particular microstate the system is in (at some instant), the larger is the value of the entropy for that thermodynamic state.

To see that this makes sense, let us review some of the ways in which the uncertainty can increase, making use of the examples in Section 6 of Chapter 1. Consider first a "loaded die" with the probabilities $P_i = \frac{1}{8}$ $(i \neq 1)$ and $P_1 = \frac{3}{8}$. A game involving this die is less uncertain than one involving a "true die," for which $P_i = \frac{1}{6}$ $(i = 1, \ldots, 6)$. We see that the more uniform the values of the P_i are, the more uncertain is the experiment. This is illustrated in Figure 4 (note that the number of events is the same in both

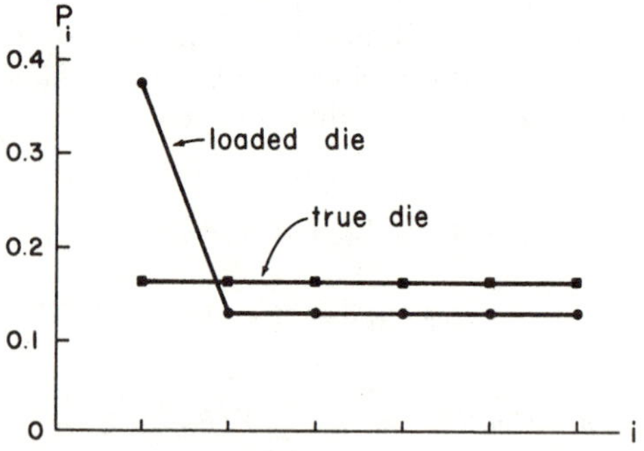

Figure 4

cases). Thus the uncertainty should increase if we go to another situation that has the same number of events but more uniform values of the probabilities. The same result also applies to the entropy. The values of the $P(E_k)$ become more uniform if β is smaller (corresponding to a higher temperature). This is illustrated in Figure 5. Therefore, at higher temperatures the probabilities of the various microstates become more uniform,

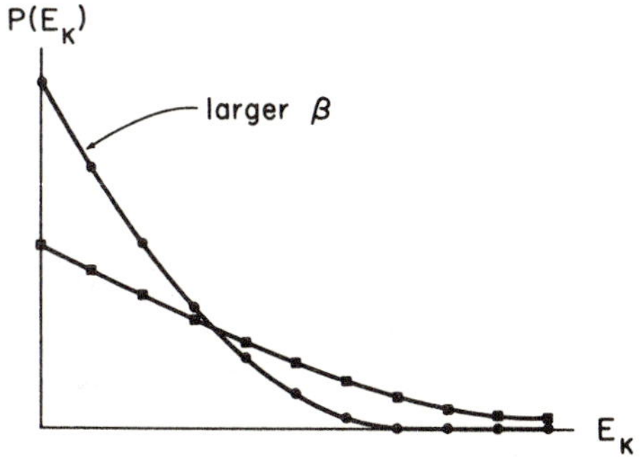

Figure 5

and hence the uncertainty (and the entropy) goes up. Physically this corresponds to the case of heating a system at a fixed volume, in which case $dS = (dQ)_{rev}/T > 0$, and the entropy indeed increases.

What happens if the volume is changed? This, as we shall now see, also produces a change in the uncertainty (or entropy). To illustrate this, again consider the case of a single true die, so that $P_i = \frac{1}{6}$ $(i = 1, \ldots, 6)$. Now if someone slips in a second true die, then the game is changed, and the probabilities become $P_{ij} = \frac{1}{36}$ $(i, j = 1, \ldots, 6)$. In both of these cases the probabilities are uniform in value — what is different is that there are *more events* in the second case. This is illustrated in Figure 6. It is clear that the uncertainty has been increased by this increase in the number of events (from ln 6 to ln 36). What is the analogous situation in the case of entropy? If the number of microstates is increased in a thermodynamic process, this

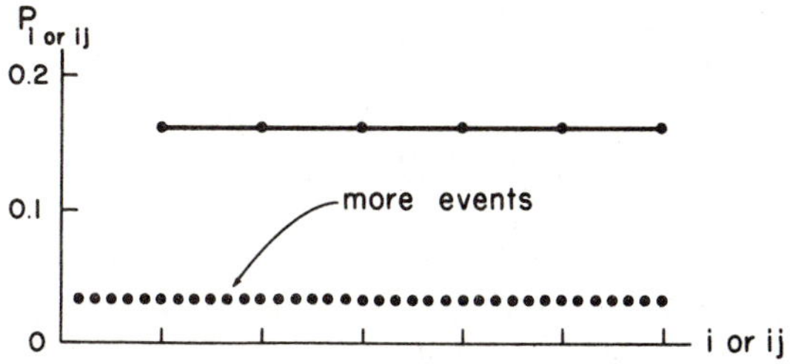

Figure 6

will then tend to increase the uncertainty — even if β is held constant. This is illustrated in Figure 7. We can conclude that any thermodynamic process that increases the number of microstates, while keeping the temperature constant, must increase the entropy.

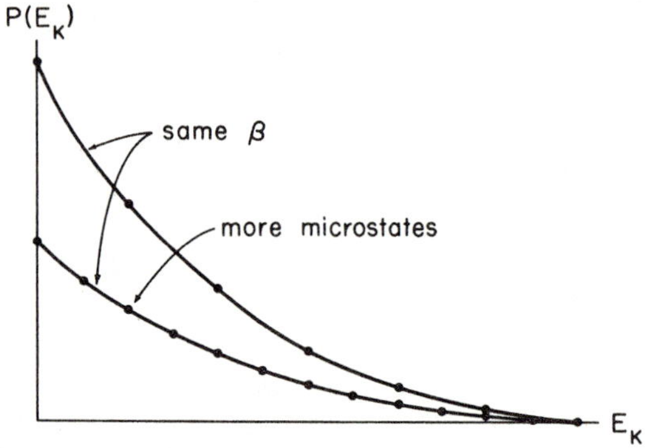

Figure 7

What is a simple way to increase the *number* of microstates? If the volume is increased, then the number of possible positions that the atoms can assume will clearly be increased. Therefore, the number of microstates (events) is increased when the volume is increased. In the case of discrete energies (shown in Figure 7) the allowed energy values become closer together when the volume is larger. (This is illustrated by ϵ_{trans} in Section 5 of Chapter 2. Also see the example in the problems.) The thermodynamic process that corresponds to an increase in the volume at constant temperature is, of course, an isothermal expansion. In such a process heat must be added to the system (because the system is doing work, and we want the temperature to remain constant). Since $(dQ)_{rev} > 0$, the entropy must also increase, which agrees with the above discussion.

In a general thermodynamic process, involving both a temperature and a volume change, the change in the uncertainty (or entropy) will depend upon the balance between the two processes described above. Thus, for example, the uncertainty increase produced by increasing the volume can be compensated by a decrease in the temperature (as in a reversible adiabatic expansion, for which $dS = 0$).

7. *FLUCTUATIONS OF THE ENERGY*

We have discussed the fact that when a system is in thermal equilibrium with a thermal reservoir, its energy is not constant but continually changes with time (i.e., fluctuates). The average value of the energy $\bar{E}$ we have already identified as being the thermodynamic internal energy U. We now ask whether we can say something about how much the energy fluctuates about the average value $\bar{E}$. As a measure of the deviation of E from $\bar{E}$ we consider the variance of E, $\text{var}(E) = \overline{(E - \bar{E})^2} = \overline{E^2} - (\bar{E})^2$. We have already noted in Section 3 that $\bar{E}$ can be expressed in terms of the partition function Z;

$$\begin{aligned} U = \bar{E} &= Z^{-1} \sum E_k P(E_k) = Z^{-1} \sum E_k\, e^{-\beta E_k} \\ &= -Z^{-1} \frac{\partial}{\partial \beta} \sum e^{-\beta E_k} = -Z^{-1} \frac{\partial Z}{\partial \beta} \end{aligned} \tag{41}$$

Now consider the term $\overline{E^2}$ contained in var (E). We have

$$\begin{aligned} \overline{E^2} &= Z^{-1} \sum E_k^2\, e^{-\beta E_k} = Z^{-1} \frac{\partial^2}{\partial \beta^2} \sum e^{-\beta E_k} \\ &= Z^{-1} \frac{\partial^2 Z}{\partial \beta^2} \end{aligned}$$

Combining this with the expression for $\bar{E}$, we obtain

$$\begin{aligned} \text{var}(E) &= \overline{E^2} - (\bar{E})^2 = Z^{-1} \frac{\partial^2 Z}{\partial \beta^2} - \left(Z^{-1} \frac{\partial Z}{\partial \beta} \right)^2 \\ &= \frac{\partial}{\partial \beta} \left(Z^{-1} \frac{\partial Z}{\partial \beta} \right) \end{aligned}$$

But this last expression for var (E) is equal to $-\partial U/\partial \beta$, as can be seen from Equation (41). Thus

$$\text{var}(E) = -\frac{\partial U}{\partial \beta} = -\frac{\partial U}{\partial T} \frac{\partial T}{\partial \beta}$$

Moreover, since $\beta = 1/kT$, $\partial \beta/\partial T = -1/(kT^2)$ or $\partial T/\partial \beta = -kT^2$. Substituting this into the last expression yields

$$\boxed{\text{var}(E) = kT^2 \left(\frac{\partial U}{\partial T} \right)_{\text{no work}}} \tag{42}$$

The expression $(\partial U/\partial T)_{\text{no work}}$ means the change in the internal energy when the temperature is changed, but no mechanical work is done on the system (e.g., constant volume, polarization, magnetization, and so on). One can see that this is correct because in the previous expressions the energies E_k in the derivative $\partial Z/\partial \beta = (\partial/\partial \beta) \Sigma\, e^{-\beta E_k}$ are treated as constant. Since the energies of the various states are constant, none of the mechanical variables, such as the volume or polarization, could be varied. Hence no work is done on the system.

In the simplest case in which the only work is $p\,dV$ work, we have $(\partial U/\partial T)_{\text{no work}} = (\partial U/\partial T) \equiv C_V$ (the heat capacity at constant volume). Thus in this case

$$\text{var}\,(E) = kT^2C_V \tag{43}$$

The standard deviation of the energy is the square root of var (E),

$$\sigma(E) = (kT^2C_V)^{1/2} \tag{44}$$

To illustrate the significance of this result, consider a perfect monatomic gas. In this case $U = \frac{3}{2}NkT$, so $C_V = \frac{3}{2}Nk$, where N is the number of atoms. Therefore the standard deviation of the energy equals

$$\sigma(E) = \left(\frac{3}{2}Nk^2T^2\right)^{1/2} = \left(\frac{3N}{2}\right)^{1/2} kT \tag{44}$$

This shows that the "distance" of the energy from the average value of the energy increases if T is increased, or if N is increased. The "distance" is, of course, in units of energy. Let us see how the standard deviation compares with the average energy by considering the ratio $\sigma(E)/\bar{E} = \sigma(E)/U$. Since $U = \frac{3}{2}NkT$,

$$\frac{\sigma(E)}{\bar{E}} = \frac{1}{(3N/2)^{1/2}} \tag{44}$$

Thus the more atoms there are in the system (the larger N), the smaller is this ratio. This means that the fluctuations of the energy are much less than $\bar{E}$ (or U) if there are many atoms in the system. If, for example, $N = 10^{18}$ (atoms),

$$\sigma(E) \simeq 10^{-9}\bar{E}$$

or one-billionth of the average energy. On the other hand if N is small, so $\sigma(E)/\bar{E}$ becomes large, the energy is not likely to be found near $\bar{E}$. Obviously if the energy is not usually found near $\bar{E}$, the internal energy ($\bar{E}$) is not a very significant property of the system. Thus, if the internal energy is to be a significant quantity (i.e., if the system is a thermodynamic system), then the system must be "macroscopic." In other words N must be large. Put another way, a system containing only a few atoms is not a thermodynamic system because the fluctuation of the energy is nearly as large as the average value of the energy. Consequently the average energy cannot be expected to represent the state of the system (its physical properties). This explains why thermodynamic systems must be macroscopic — in the sense that N must be large.

8. *OPEN SYSTEMS*

Up to this point we have considered only those systems which have a fixed number of particles (closed systems). In many cases it is necessary, or at least very convenient, to consider systems in which the number of particles may vary (open systems). This is accomplished by allowing the system to exchange particles with another system (which can be thought of as a "particle reservoir"). An example of such a situation is a liquid in equilibrium with its vapor. In this case the number of atoms in the liquid, as well as the energy of the liquid, is continually changing. Thus an open system is one that can exchange both particles (with a particle reservoir) and energy (with a thermal reservoir — which may be distinct from the particle reservoir). The exchange of these two quantities is quite independent, and hence the energy E and the number of particles N are independent variables of the system.

The microstates of an open system are defined in the same way as for closed systems, except that now a system is also considered to be in a different microstate if the value of N is changed. The same reasoning as before then leads to the natural extension of our previous *basic assumption*, namely:

> All (micro)states with the same energy E and the same number of particles N are assumed to be equally probable. (45)

The probability of a particular (micro)state is now represented by $P(E, N)$. In order to determine how $P(E, N)$ depends on the variables E and N, we use a method analogous to the one used for closed systems (Section 2).

We begin by considering a composite system $A + B$ made up of the macroscopic systems A and B. These systems are free to exchange both energy and particles with one another, and also with a common thermal and particle reservoir. The probability that the composite system is in a state with the energy E_{A+B} and N particles will be represented by $P_{A+B}(E_{A+B}, N)$. The total number of particles is $N = N_A + N_B$, where N_A and N_B are the number of particles in systems A and B, respectively. Moreover, since A and B are macroscopic systems, $E_{A+B} \simeq E_A + E_B$. Therefore $P_{A+B}(E_{A+B}, N) = P_{A+B}(E_A + E_B, N_A + N_B)$. As in the case of closed systems, we again argue that the energies of the two systems are independent of one another. Moreover, by the same reasoning, the number of particles in one system is independent of the number in the other system (since they are both free to exchange particles with the particle reservoir). Hence we conclude that the microstate (E_A, N_A) is independent of (E_B, N_B), so that

$$P_{A+B}(E_A + E_B, N_A + N_B) = P_A(E_A, N_A)P_B(E_B, N_B) \tag{46}$$

To determine how $P(E, N)$ depends on E and N, we now use the same method as in Section 2. We note that the derivative of the left side of Equation (46) with respect to E_A equals the derivative with respect to E_B, hence the right sides must be equal. Therefore

$$P_B(E_B, N_B)\frac{\partial P(E_A, N_A)}{\partial E_A} = P_A(E_A, N_A)\frac{\partial P_B(E_B, N_B)}{\partial E_B} \tag{47}$$

Moreover, if we differentiate Equation (46) with respect to N_A or N_B, the right side yields

$$P_B(E_B, N_B)\frac{\partial P_A(E_A, N_A)}{\partial N_A} = P_A(E_A, N_A)\frac{\partial P_B(E_B, N_B)}{\partial N_B} \tag{48}$$

Considering Equation (47) first, we have

$$\frac{\partial \ln P_A(E_A, N_A)}{\partial E_A} = \frac{\partial \ln P_B(E_B, N_B)}{\partial E_B}$$

and hence each side must equal a constant (the same constant), which we again call $-\beta$. Integrating, we have

$$\ln P_A(E_A, N_A) = -\beta E_A + C_A(N_A)$$

where $C_A(N_A)$ is an unknown function of N_A. Thus

$$P_A(E_A, N_A) = e^{-\beta E_A}\, e^{C_A(N_A)} \tag{49}$$

and, of course, we get a similar result for system B. To determine the function $C(N)$, we now turn to Equation (48). From it we find that

$$\frac{\partial \ln P_A(E_A, N_A)}{\partial N_A} = \frac{\partial \ln P_B(E_B, N_B)}{\partial N_B}$$

Once again, the left- and right-hand sides depend on different independent variables, so they must both be equal to a constant (the same constant), which we shall call $\beta\mu$. The constant μ, like β, does not depend on the composition of the system. It must characterize the *particle* reservoir, in the same way that β represents the thermal reservoir. We have therefore

$$\frac{\partial \ln P_A(E_A, N_A)}{\partial N_A} = \beta\mu$$

Substituting the result (49) into this equation, we find that

$$\frac{dC_A(N_A)}{dN_A} = \beta\mu$$

or

$$C_A(N_A) = \beta\mu N_A + D_A$$

where D is a constant (independent of E and N). Substituting this result back into Equation (49), we obtain

$$P_A(E_A, N_A) = e^{D_A}\, e^{-\beta E_A + \beta\mu N_A}$$

Therefore the probability that the system is in a particular state, with energy E and N particles, is given by (dropping the subscript A)

$$\boxed{P(E, N) = \mathcal{Z}^{-1} e^{-\beta E + \beta \mu N}} \tag{50}$$

where we have set $\mathcal{Z}^{-1} = e^{D}$. (The energy E is, of course, the energy of the system of N particles.) The constant $\mathcal{Z}$ is now the new partition function and has the value determined by the normalization condition

$$\sum_{N=0}^{\infty} \sum_{k} P(E_k, N) = 1$$

Using (50), we have

$$\boxed{\mathcal{Z} = \sum_{N=0}^{\infty} \sum_{k} e^{+\beta \mu N - \beta E_k} \equiv \sum_{N=0}^{\infty} e^{\beta \mu N} Z_N} \tag{51}$$

where the energy sum is over all the states k (energy E_k) of the system with N particles, including any repetition due to the fact that different states have the same energy (degenerate states). This sum is then followed by a second sum over all allowed values of N. Since the first sum $\sum_k e^{-\beta E_k}$ is simply the partition function for a system with N particles (which we now denote by Z_N), the partition function $\mathcal{Z}$ can also be written in the second form of Equation (51).

It should perhaps be noted that the present results have nothing to do with quantum mechanics per se. The results (50) and (51) are equally valid for the classical case where the energy can vary continuously, except that one must use a distribution function. That is, one has

$$f[E(\mathbf{r}_1, \ldots, \mathbf{v}_N), N] = \mathcal{Z}^{-1} e^{+\beta \mu N - \beta E(\mathbf{r}_1, \ldots, \mathbf{v}_N)} \tag{52}$$

and

$$\mathcal{Z} = \sum_{N=0}^{\infty} e^{+\beta \mu N} \int_V \cdots \int_{-\infty}^{\infty} e^{-\beta E(\mathbf{r}_1, \ldots, \mathbf{v}_N)} \, d^3 r_1 \cdots d^3 v_N \tag{53}$$

The distributions (50) and (52) are known as **"grand canonical" distributions** and $\mathcal{Z}$ is called the **grand canonical partition function.**

Our interest in using open systems will be to help us treat certain quantum-mechanical cases, so we shall be primarily interested in the results (50) and (51). The advantage of using open systems rather than closed systems with fixed values of N will become clear in Chapter 4. It is essentially a device that makes calculations much easier, and it has nothing to do with whether or not the system is treated classically or quantum-mechanically.

The thermodynamic state functions can again be expressed in terms of the partition function Z, just as in the case of closed systems. The only difference in the present case is that N can vary. For either an open or a closed system the combined first and second law of thermodynamics can be written

$$du = T\,ds - p\,dv$$

where u, s, and v are all specific (intensive) quantities. (For example, u is the internal energy per particle.) If we let $\bar{N}$ represent the (average) number of particles in the system, then the *total* internal energy, entropy, and volume is

$$U = \bar{N}u, \qquad S = \bar{N}s, \qquad V = \bar{N}v$$

and the above equation can be written in the form

$$dU - u\,d\bar{N} = T(dS - s\,d\bar{N}) - p(dV - v\,d\bar{N})$$

or, collecting the terms $d\bar{N}$,

$$dU = T\,dS - p\,dV + (u - Ts + pv)\,d\bar{N} \tag{54}$$

The quantity $(u - Ts + pv)$ is simply the Gibbs free energy per particle.

We shall now show that the new constant μ which appears in the partition function (51), is just the Gibbs free energy per particle. To do this, we consider the change in $(\ln Z)$

$$d(\ln Z) = \frac{\partial}{\partial \beta}(\ln Z)\,d\beta + \frac{\partial}{\partial V}(\ln Z) + \frac{\partial}{\partial \mu}(\ln Z)\,d\mu \tag{55}$$

Now

$$\frac{\partial}{\partial \mu}(\ln Z) = Z^{-1}\frac{\partial}{\partial \mu}\sum_{N=0}^{\infty}\sum_{k} e^{\beta\mu N - \beta E_k} = Z^{-1}\sum_{N=0}^{\infty}\sum_{k}\beta N\, e^{\beta\mu N - \beta E_k}$$

But the average number of particles $\bar{N}$ is defined by

$$\bar{N} \equiv \sum_{N=0}^{\infty}\sum_{k} NP(E_k, N) = Z^{-1}\sum_{N=0}^{\infty}\sum_{k} N\, e^{\beta\mu N - \beta E_k} \tag{56}$$

So we find that

$$\frac{\partial}{\partial \mu}(\ln Z) = \beta\bar{N} \tag{57}$$

Also

$$\frac{\partial}{\partial \beta}(\ln Z) = Z^{-1}\sum_{N=0}^{\infty}\sum_{k}(-E_k + \mu N)\, e^{\beta\mu N - \beta E_k} = (-\bar{E} + \mu\bar{N}) \tag{58}$$

Substituting (57) and (58) into (55), we obtain

$$d(\ln Z) = (-\bar{E} + \mu\bar{N})\,d\beta + \frac{\partial}{\partial V}(\ln Z)\,dV + \beta\bar{N}\,d\mu$$

$$= d[\beta(-E + \mu\bar{N})] + \beta d(\bar{E} - \underline{\mu\bar{N}}) + \frac{\partial}{\partial V}(\ln Z)\, dV$$
$$+ \underline{\beta d(\mu\bar{N})} - \beta\mu\, d\bar{N}$$

We see that the underlined terms cancel, and taking the first differential over to the other side and multiplying by β^{-1}, we obtain

$$\beta^{-1}d[\ln Z + \beta(\bar{E} - \mu\bar{N})] = d\bar{E} + \beta^{-1}\frac{\partial}{\partial V}(\ln Z)\, dV - \mu\, d\bar{N} \tag{59}$$

Comparing this result with Equation (54),

$$T\, dS = dU + p\, dV - (u - Ts + pv)\, d\bar{N} \tag{54}$$

we conclude (using the fact that $\beta = 1/kT$) that the thermodynamic quantities u, s, g, and p are related to the partition function by

$$U = \bar{E} \tag{60}$$
$$S = k[\ln Z + \beta\bar{E} - \beta\mu\bar{N}] \tag{61}$$
$$p = kT\frac{\partial}{\partial V}(\ln Z) \tag{62}$$
$$g \equiv u - Ts + pv = \mu \tag{63}$$

Also, from Equations (57) and (58)

$$\bar{N} = kT\frac{\partial}{\partial \mu}(\ln Z) \tag{64}$$
$$\bar{E} = -\frac{\partial}{\partial \beta}(\ln Z) + \mu kT\frac{\partial}{\partial \mu}(\ln Z) \tag{65}$$

Using these expressions, one can easily obtain expressions for the remaining state functions of thermodynamics.

A particularly interesting result is obtained by noting that

$$pV = \bar{N}(u - Ts + pv) - \bar{N}(u - Ts)$$
$$= \mu\bar{N} - (\bar{E} - kT\ln Z - \bar{E} + \mu\bar{N}) = kT\ln Z$$

or, in other words,

$$Z = e^{pV/kT} = e^{\beta pV} \tag{67}$$

Thus, the grand canonical distributions (50) and (52) can be written

$$P(E, N) = e^{-\beta(E-\mu N+pV)} \tag{68}$$
$$f[E(\mathbf{r}_1, \ldots, \mathbf{v}_N), N] = e^{-\beta(E(\mathbf{r}_1,\ldots,\mathbf{v}_N)-\mu N+pV)}$$

The expressions (67) and (68) may be compared with (38) for the case of closed systems.

ESSENTIAL POINTS

1. A microstate is a detailed microscopic description of the state (or condition) of a system to which we can assign a probability. Classically a microstate is specified by the values of $(\mathbf{r}_1, \ldots, \mathbf{v}_N)$ within an accuracy $(d^3r_1, \ldots, d^3v_N)$. Quantum-mechanically it is specified by the values of appropriate quantum numbers (which depend on the physical system).
2. The basic assumption of statistical mechanics is that all microstates that have the same energy are equally probable.
3. For a closed system (fixed value of N) in thermal equilibrium, the classical distribution function is

$$f[E(\mathbf{r}_1, \ldots, \mathbf{v}_N)] = D_N Z^{-1} e^{-\beta E}$$

where the partition function (normalization factor) is given by

$$Z = D_N \int_V d^3r_1 \cdots \int_{-\infty}^{\infty} d^3v_N \, e^{-\beta E}$$

and we can choose $D_N = 1$. The quantum expression is

$$P(E_k) = Z^{-1} e^{-\beta E_k}, \qquad \text{where} \qquad Z = \sum_k e^{-\beta E_k}$$

and k represents the appropriate quantum numbers. The quantities $f(E)\, d^3r_1 \cdots d^3v_N$ and $P(E_k)$ represent the probability of a *particular* microstate. An energy level is the collection of all microstates with the same energy (a compound event).

4. The thermodynamic internal energy U is identified as being the average energy in a system, which can be expressed as $U = -\partial \ln Z/\partial\beta$. The parameter β is then found to be equal to $(1/kT)$, where k is the Boltzmann constant. Comparing $d(\ln Z)$ with the second law of thermodynamics leads to the identification of the entropy and equation of state in terms of Z:

$$dS = k\, d(\ln Z + \beta U), \qquad p = -kT\left(\frac{\partial \ln Z}{\partial V}\right)$$

5. The entropy of a thermodynamic state is equal to Boltzmann's constant times the statistical uncertainty of that thermodynamic state:

$$S = +kH = -k \sum_k P(E_k) \ln P(E_k)$$

PROBLEMS

1. A hypothetical system has only two (micro)states, represented by $\uparrow$ and $\downarrow$. The energy of the system when it is in these states is $E(\uparrow) = -1$ and $E(\downarrow) = +1$ (in suitable units of energy). Assume that, when this system is in thermal equilibrium with a reservoir, the probability that the system is in three states is found to be

$$P(\uparrow) = 0.7, \qquad P(\downarrow) = 0.3$$

(a) Determine the constants C and β in $P(E) = Ce^{-\beta E}$.

(b) Consider a composite system made up of two of these (independent) systems. The composite system is coupled to the same reservoir. Complete the following table for the remaining states.

State	*Energy*	*Probability of State*
$\uparrow\uparrow$	-2	0.49

(c) What is the probability that this composite system has zero energy?

(d) Show that the probabilities of the states in part (b) satisfy

$$P(E) = De^{-\beta' E}$$

Determine the relationship between (D, β') and (C, β), which were obtained in part (a). Comment on which of these constants is dependent on (or independent of) the system.

(e) Determine the value of U for the composite system.

2. A simple system has different microstates with energies

$$-1, 0, 0, 0, +1 \qquad (\times 10^{-20} \text{ joule})$$

(a) Determine the probability that the system is in these different states, if $T = 200°\text{K}, 400°\text{K}$.

(b) In each case what is the probability that the energy of the system is zero?

(c) What is the relative chance of finding the system in a state with $E = -1 \times 10^{-20}$ joule compared to the chance of finding the system in a state with $E = +1 \times 10^{-20}$ joules? Set up a general expression for the relative probability of finding a system in a state with $E = E_1$, and with $E = E_2$ (in terms of β and Z).

3. A hypothetical system contains three different molecules, each of which is capable of having the energies

$$\epsilon_i = 0, 1, 2, 3, \ldots \qquad (\times 2.4 \times 10^{-14} \text{ erg})$$

Assume that there is no interaction energy between the molecules so that the total energy of the system is $E = \epsilon_1 + \epsilon_2 + \epsilon_3$.

(a) Some of the values of $\Omega(E)$ — the number of microstates of energy E — are listed in the table. Complete this table.

$E(2.4 \times 10^{-14}$ erg)	0	1	2	3	4	5	6	7	8	9	10
$\Omega(E)$			6				28	36	45	55	66

(b) Determine the values of $e^{-\beta E}$ for the values of E in (a), if $T = 290°K$.

(c) From the results in (b) determine the value of the partition function (using the result in the example in the text).

(d) Plot the probabilities of the microstates and the energy levels as a function of E. What is the most probable value of E for this system?

4. Show that if we redefine the energies of all microstates E_k so that the new values are given by $E'_k = E_k + E^0$ (where E^0 is some constant), then the new internal energy

$$U' = (Z')^{-1} \sum E'_k e^{-\beta E'_k}$$

is simply related to the internal energy U (computed from the values E). Determine this relationship. Does this change in the value of U have any physical significance?

5. Show that Equation (23) holds for any classical system that has a total energy of the form

$$E = \sum_{i=1}^{N} \epsilon(\mathbf{r}_i, \mathbf{v}_i)$$

where $\epsilon(\mathbf{r}, \mathbf{v})$ is any function of $(\mathbf{r}, \mathbf{v})$. Obtain an expression for the particle partition function in this case.

6. In the identification of β, the internal energy was taken to be $U = 3NkT/2$. It might be argued that one could equally well take $U = 3Nk(T + T_0)/2$ (T_0 being some constant), since this yields the same value for $C_V = (\partial U/\partial T)_V$. Show that in this case the resulting β would yield a pressure for a perfect gas, $p = \beta^{-1} (\partial \ln Z/\partial V)_T$, such that p/T would depend on T (unless $T_0 = 0$). This result is not acceptable, because the *absolute temperature scale* is *defined* to be proportional to p for a sufficiently rarefied gas (if V is constant).

7. From a practical viewpoint, the principal difficulty in statistical mechanics is to obtain the partition function as an explicit function of β and V. Assume, however, that some clever person obtained the following approximate partition functions. In each case determine the equation of state and the heat capacity C_V of the system, and from these facts identify the hypothetical systems

(a) $Z = V^N(2\pi/\beta M)^{5N/2}$.

(b) $Z = (V - Nb)^N(2\pi/\beta M)^{3N/2} e^{\beta N^2/V}$ (a, b: constants).

8. Use the partition function for a perfect monatomic gas, and the equation

$$S = k(\ln Z + \beta U) + C(N)$$

(a) Determine how the "constant" $C(N)$ must depend on N if one makes the *additional* requirement

$$S(2V, 2N) = 2S(V, N)$$

This requires that the entropy of a system will double if the size (volume and mass) is doubled. A more detailed discussion is given in Appendix D.

(b) Show that the requirement in part (a) can also be satisfied if one takes

$$S = k(\ln Z + \beta U)$$

but one uses a different convention than (14) for D_N. How would D_N have to depend on N in this case in order to satisfy the requirement in part (a)?

9. (a) Show that the Gibbs function $G = U - TS + pV$ is given by $G = kTV^2\,\partial(V^{-1}\ln Z)/\partial V$.

(b) Show that the enthalpy, $H = U + pV$, is related to the partition function by $H = -(\partial \ln Z/\partial\beta) + kTV(\partial \ln Z/\partial V)$.

10. Using the partition function for a perfect monatomic gas, obtain an expression for the entropy. Write S in the form of the sum of terms, one involving V and the other containing T. Give a brief *statistical interpretation* for the variation in S when each of these terms is varied.

11. The energies of the microstates for a quantum-mechanical perfect monatomic gas depend on the volume according to $E_k = A_k V^{-2/3}$, where the A^k do not depend on the volume [this is due to the fact that the quantum translational energy ϵ_{trans} [Equation (40), Chapter 2] is proportional to $L^{-2} = (V^{1/3})^2$]

(a) Sketch a figure similar to Figure 7 for two equilibrium states $V = V_0$ and $V = 8V_0$ (same β), indicating carefully the change in the density of the microstates. Why are all $P(E_k)$ smaller in the second equilibrium state? Which thermodynamic state has the larger entropy?

(b) If the uncertainty (entropy) is the same for two equilibrium states (β_0, V_0) and $(\beta_1, 8V_0)$, how must β_1 be related to β_0? [Hint: Make the probabilities of microstates with different energies equal.] Draw a figure similar to Figures 5 and 7 for the *two* equilibrium states, indicating which states have the same probability.

12. (a) Show that the entropy, when expressed in terms of the probability of the energy levels $P_\ell\,(E)$, can be written in the form

$$S = k \sum_{\text{levels}} P_\ell(E) \ln \Omega(E) - k \sum_{\text{levels}} P_\ell(E) \ln P_\ell(E)$$

(b) In the case of a microcanonical ensemble (for which the energy has a definite value, say $E = E_0$), what is the resulting expression for S? (A similar expression for S occupies a prominent position on the gravestone of L. Boltzmann. This indicates the importance that some people place on the statistical interpretation of entropy!) Note that as $x \to 0$, $x \ln x$ also goes to zero.

(c) This system now undergoes a free (i.e., no work) adiabatic expansion, so that its energy is unchanged. Discuss briefly what happens to the value of the degeneracy $\Omega(E)$. From the result of part (b), what can you conclude about the change in the entropy?

13. Copper has an atomic weight of 63.5 and a density of 8.93 gm/cm^3. Assume that its heat capacity obeys the Dulong-Petit law, $C_V = 3Nk$. Consider a very small sample of copper, of 10^{-3} cm^3, that is in equilibrium at 1000°C. Determine the standard deviation of the energy of this system and compare it with its internal energy (taken to be zero at $T = 0°K$). If an amount of energy equal to $\sigma(E)$ were used to raise this system in a gravitational field, what would be the displacement?

14. Using an analysis similar to the one used for computing var(E), show that for open systems

$$\text{var}\,(N) = \beta^{-1}\,(\partial \bar{N}/\partial \mu)_{V,T}$$

15. Show that the entropy for an open system, Equation (61), can also be expressed in a form similar to Equation (39), where now the probabilities are given by Equation (50).

4

Applications of Statistical Mechanics

1. THE PROBLEM

In the last chapter we obtained a theory, based on certain statistical assumptions, that can be used to predict the physical properties of various systems, and even the probability that the system is in a particular microstate $f[E(\mathbf{r}_1, \ldots, \mathbf{v}_N)]\, d^3r_1 \cdots d^3v_N$ [or $P(E_K) = Z^{-1} e^{-\beta E_K}$]. In principle all we need to do is to obtain an expression for the energy of the system $E(\mathbf{r}_1, \ldots, \mathbf{v}_N)$ (or its eigenvalues E_K), and then evaluate the partition function

$$Z = \int_V \cdots \int_{-\infty}^{\infty} e^{-\beta E}\, d^3r_1 \cdots d^3v_N \qquad \left(\text{or} \sum e^{-\beta E_K}\right) \tag{1.1}$$

Once this is done, we can predict the distribution function for that system, and all its thermodynamic properties, using relationships such as

$$U = -\left(\frac{\partial \ln Z}{\partial \beta}\right)_V \quad \text{or} \quad p = -kT\left(\frac{\partial \ln Z}{\partial V}\right)_\beta \quad \text{(equation of state)} \tag{1.2}$$

It might appear, therefore, that the applications of statistical mechanics should be a relatively simple matter: namely, the evaluation of the partition function for the system in question. At this point one might reasonably ask, "What is the problem?" The problem is concealed in what is meant by "evaluate" the partition function. As it stands, the expression for the partition function, Equation (1.1), is of no use in Equation (1.2) because the dependency of Z on V and β is far from obvious. Before equations of the type (1.2) can be used, Z must be determined as an *explicit* function of V and β. In other words, to "evaluate" the partition function, we must actually perform the integrations (or summations) and obtain $Z(\beta, V)$ as a function of β and V. For complicated functions of the energy $E(\mathbf{r}_1, \ldots, \mathbf{v}_N)$ this evaluation is essentially impossible because the integrals cannot be performed in closed form. In this case we must either select a simpler function for the energy or revert to a numerical integration of the partition function (e.g., using an electronic computer). In practice even this latter possibility has severe limitations because of the enormous number of integration variables. Thus we are usually led to the problem of selecting a *model* for our physical system that has a sufficiently simple energy to make the evaluation of Z possible. Ideally a model should serve two purposes:

(a) It should have a simple enough expression for the energy of the system for $Z(\beta, V)$ to be obtained as an explicit function of (β, V).

(b) It should be realistic enough for the resulting $Z(\beta, V)$ to predict the experimentally observed physical properties of the system with "reasonable" accuracy.

These criteria of simplicity and realism cannot be satisfied for all physical systems — after all, physical systems are not necessarily simple! Thus, though there are models for such systems as ferromagnets, liquids, and dense gases, they do not satisfy one or the other of the criteria above. In this case a more elaborate analysis must be used, and many of the properties of such systems are not yet entirely understood. In the present chapter we shall consider systems for which we can obtain models that satisfy both (a) and (b). A number of such models were discussed in Chapter 2. Sometimes we may even have two possible models for one system, both of which satisfy condition (a). By examining which of these models are realistic [in the sense of (b)], we can learn to identify which molecular (microscopic) properties are responsible for the various thermodynamic (macroscopic) properties of a system. In this way one can acquire an insight, or "feel," about the molecular origins of the physical properties of different systems.

In this chapter we shall begin by considering a classical perfect gas, which has already been discussed briefly in Chapter 3. We shall then proceed to consider other systems and illustrate the need for quantum statistical mechanics by showing that the classical models do not always satisfy (b). The conditions under which the classical models are an accurate description of physical systems will be considered in Section 6 and illustrated in subsequent sections.

2. MAXWELLIAN DISTRIBUTION FUNCTIONS

In this section we shall consider the statistical descriptions of the *translational motion* of molecules in a perfect gas. Several applications of these results to specific problems will be considered in Section 3. In the present section we simply wish to discuss several distribution functions that describe various aspects of this translational motion. For our present purposes we can use classical mechanics to describe the translational motion of the molecules (it will be shown in Sections 9 and 15 that this is justifiable except in the cases of extremely low temperatures or very high densities). Moreover, since we are only concerned at present with the *translational* motion of the molecules (and not their rotational or vibrational motion), we can simply ignore the structure of the molecules. That is, we can effectively picture the molecule as a monatomic molecule having mass m (the total mass of the molecule) and a velocity $\mathbf{v}$ (the velocity of the center of mass). With this understanding, we can now use the results that were obtained in the last chapter for a monatomic gas (Section 4). It was shown there that if

$f(\epsilon)\, d^3r\, d^3v =$ probability of finding a molecule in the region (d^3r, d^3v) about $(\mathbf{r}, \mathbf{v})$

then

$$f(\epsilon) = V^{-1}\left(\frac{m}{2\pi kT}\right)^{3/2} \exp\left(-\frac{m\mathbf{v}^2}{2kT}\right) \tag{2.1}$$

where V is the volume of the system, $\epsilon = \frac{1}{2}m\mathbf{v}^2$ (the translational or kinetic energy of the molecule), and $k = 1.38 \times 10^{-16}$ ergs/°K is Boltzmann's constant. Usually we are not interested in the probability of a *particular* molecule, but rather the *probable number* of molecules in the region (d^3r, d^3v) of $(\mathbf{r}, \mathbf{v})$. We obtain this quantity by multiplying $f(\epsilon)\, d^3r\, d^3v$ by the total number of molecules, N. Therefore, if we set (η = Greek eta, for "number")

$\eta(\mathbf{r}, \mathbf{v})\, d^3r\, d^3v =$ probable number of molecules in the region (d^3r, d^3v) of $(\mathbf{r}, \mathbf{v})$ (2.2)

then, by multiplying Equation (2.1) by N, we conclude that

$$\eta(\mathbf{r}, \mathbf{v}) = \left(\frac{N}{V}\right)\left(\frac{m}{2\pi kT}\right)^{3/2} \exp\left(-\frac{m\mathbf{v}^2}{2kT}\right) \tag{2.3}$$

This distribution function is normalized to N, whereas $f(\epsilon)$ is normalized to unity.

The first point to be noted about these distributions is that they are independent of $\mathbf{r}$. This means that the molecules have an equal chance of being found in the region d^3r of *any* point within the system. This makes sense, for we do not expect to find a density gradient in this system when it is in thermal equilibrium. Thus the spatial variable is really superfluous, and we can consider the probable number of molecules *in the whole system* that have velocities in the range d^3v of $\mathbf{v}$. This quantity we shall denote by $\eta(\mathbf{v})\, d^3v$. It is given by integrating Equation (2.3) over all points in the volume V, or

$$\eta(\mathbf{v}) = \int_V \eta(\mathbf{r}, \mathbf{v})\, d^3r = V\eta(\mathbf{r}, \mathbf{v})$$

so

$$\eta(\mathbf{v}) = N\left(\frac{m}{2\pi kT}\right)^{3/2} \exp\left(-\frac{m\mathbf{v}^2}{2kT}\right) \tag{2.4}$$

where now

$$\eta(\mathbf{v})\, d^3v = \begin{array}{l}\text{the probable number of molecules with velocities}\\ \text{in the range } d^3v \equiv dv_x\, dv_y\, dv_z \text{ of } \mathbf{v}\end{array} \tag{2.5}$$

The distribution function $\eta(\mathbf{v})$ is called the **Maxwellian velocity distribution** function in honor of J. C. Maxwell, who first obtained it in 1857.

Now let us examine some of the features of this distribution function and note again what it means. If we draw a set of coordinate axes (v_x, v_y, v_z), as shown in Figure 2.1, we can represent the velocity of a molecule by a single point. If we put a point in this space for every molecule, we then obtain a "cloud" of points. These points move around in some erratic fashion,

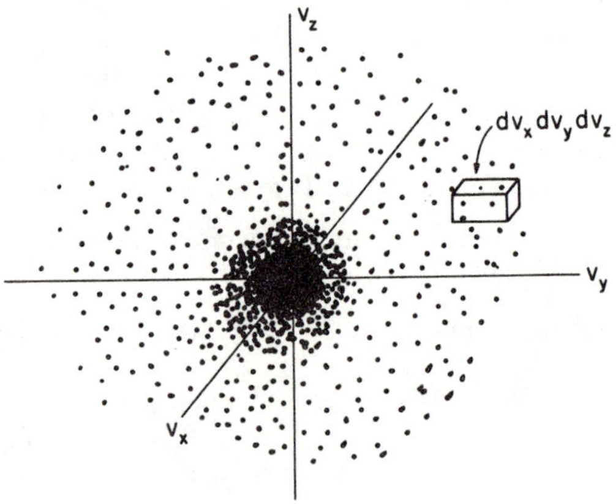

Figure 2.1

changing their location every time the velocity of a molecule is altered (because of a collision with the walls or another molecule). A snapshot (at some instant of time) might yield the picture in Figure 2.1. The distribution function $\eta(\mathbf{v})$ tells us the *probable* number of molecules we can expect to find in a region $dv_x\, dv_y\, dv_z$ about $\mathbf{v}$. That is, $\eta(\mathbf{v})\, d^3v$ gives the *average number* found in this region in a *large number* of such snapshots (experiments). It does *not* represent the actual number of molecules found in any *particular* snapshot (obviously this number would vary from snapshot to snapshot). Equation (2.4) shows that the probable number depends only on $\mathbf{v}^2$, and this means that, on the average, the points in Figure 2.1 will be symmetrically distributed about the origin $\mathbf{v} = 0$. Moreover, since the factor $\exp(-m\mathbf{v}^2/2kT)$ decreases as $\mathbf{v}^2$ increases, the molecules can be expected to be predominantly located near the origin of Figure 2.1. We shall see shortly how different temperatures affect the distribution of velocities.

In some situations we may want to know the probable number of molecules that have a certain value of some *component* of the velocity. Thus, we might want to know (for example) the probable number of molecules with a y component of the velocity in the range dv_y of v_y. In terms of our velocity space, we are asking for the probable number of points between v_y and $v_y + dv_y$ (see Figure 2.2). This number is obtained by adding up the prob-

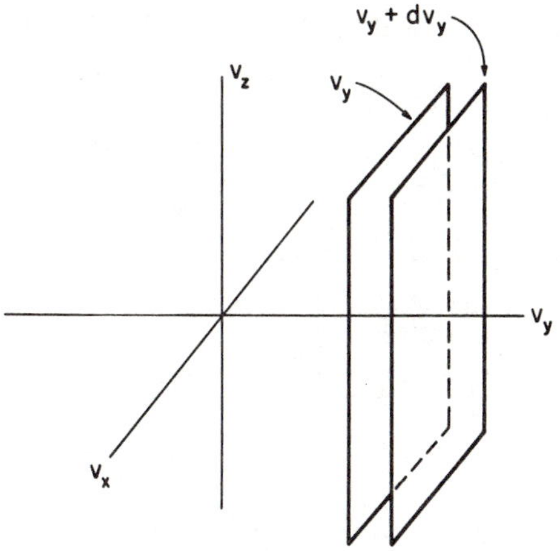

Figure 2.2

able number of molecules between the two (infinite) planes indicated in Figure 2.2. Thus if we set

$$\eta(v_y)\, dv_y = \begin{array}{l}\text{the probable number of molecules with a } y \text{ com-}\\ \text{ponent of the velocity in the range } dv_y \text{ of } v_y\end{array} \tag{2.6}$$

then, using Equation (2.4) for $\eta(\mathbf{v})$, we find

$$\eta(v_y)\,dv_y = dv_y \int_{-\infty}^{\infty} dv_x \int_{-\infty}^{\infty} dv_z\,\eta(\mathbf{v}) = N\left(\frac{m}{2\pi kT}\right)^{1/2} \exp\left(-\frac{mv_y^2}{2kT}\right) dv_y \cdot$$

$$\int_{-\infty}^{\infty} dv_x \left(\frac{m}{2\pi kT}\right)^{1/2} \exp\left(-\frac{mv_x^2}{2kT}\right) \int_{-\infty}^{\infty} dv_z \left(\frac{m}{2\pi kT}\right)^{1/2} \exp\left(-\frac{mv_z^2}{2kT}\right)$$

Using the Gaussian integrals in Appendix A, we find that each integral equals unity, so

$$\eta(v_y) = N\left(\frac{m}{2\pi kT}\right)^{1/2} \exp\left(-\frac{mv_y^2}{2kT}\right) \tag{2.7}$$

Obviously, a similar result holds for the x and z components of the velocity [by replacing v_y in Equation (2.7) with v_x or v_z].

If $\eta(v_x)$ is plotted as a function of v_x, one obtains the result shown in Figure 2.3. Two cases are shown, corresponding to two different tempera-

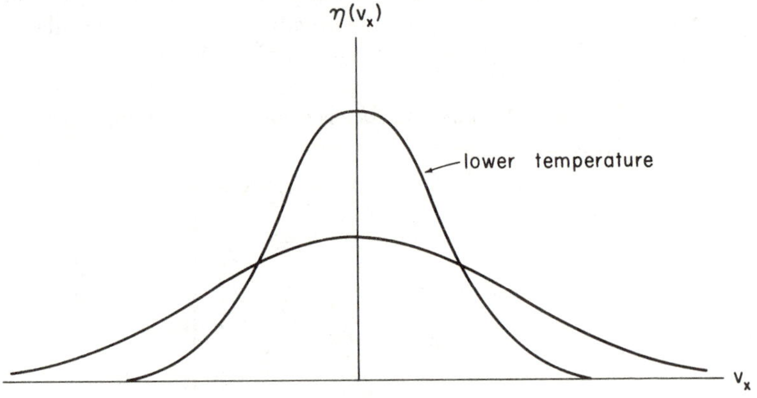

Figure 2.3

tures. The area under both curves must be the same, for η is normalized to N,

$$\int_{-\infty}^{\infty} \eta(v_x)\,dv_x = N$$

[as can be shown from Equations (2.7) and (A.1)]. This, of course, is simply an expression of the fact that all atoms have *some* velocity v_x. It can be seen from Figure 2.3 that there are the same number of molecules moving in the positive and negative x directions. This is simply a consequence of the fact that $\eta(v_x)$ depends only on v_x^2. It is the same symmetry (about the origin) that was noted in Figure 2.1. This means that the average value of v_x in this system is zero, for

$$\bar{v}_x = \int_{-\infty}^{\infty} v_x \eta(v_x)\,dv_x$$

and the integrand is an odd function of v_x, and therefore the integral must vanish.

Another distribution function, besides $\eta(\mathbf{v})$ and $\eta(v_x)$, is also of interest. This distribution function tells us the probable number of molecules that are in the region dv of a certain *speed* v. Let

$$\eta(v)\,dv = \begin{array}{l}\text{probable number of molecules in the system with a}\\ \text{speed in the range } dv \text{ of } v\end{array} \tag{2.8}$$

The speed v is related to the velocity $\mathbf{v}$ by

$$v = [v_x^2 + v_y^2 + v_z^2]^{1/2} \tag{2.9}$$

$\eta(v)$ can readily be obtained from $\eta(\mathbf{v})$. Referring to Figure 2.4, we see that the speed v is simply the distance from the origin in this velocity space. According to Equation (2.8), $\eta(v)\,dv$ is the probable number of molecules in

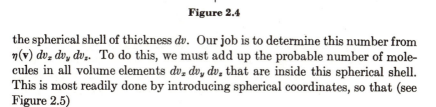

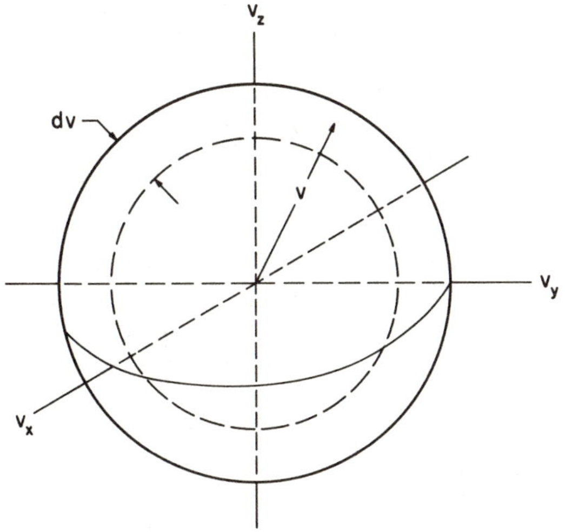

Figure 2.4

the spherical shell of thickness dv. Our job is to determine this number from $\eta(\mathbf{v})\,dv_x\,dv_y\,dv_z$. To do this, we must add up the probable number of molecules in all volume elements $dv_x\,dv_y\,dv_z$ that are inside this spherical shell. This is most readily done by introducing spherical coordinates, so that (see Figure 2.5)

$$dv_x\,dv_y\,dv_z = v^2\,dv \sin\theta\,d\theta\,d\phi$$

Then

$$\eta(v)\,dv = \int_0^{2\pi} d\phi \int_0^{\pi} d\theta\,\eta(\mathbf{v}) \sin\theta\, v^2\,dv$$

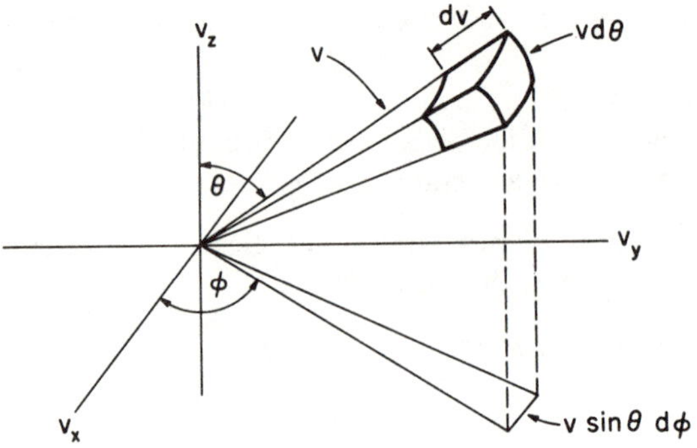

Figure 2.5

Using Equation (2.4), we see that $\eta(\mathbf{v})$ depends only on v^2 and hence the angular integrals can be performed, giving 4π. So

$$\eta(v)\,dv = 4\pi\eta(\mathbf{v})v^2\,dv$$

or

$$\eta(v) = N\sqrt{\frac{2}{\pi}}\left(\frac{m}{kT}\right)^{3/2}\exp\left(-\frac{mv^2}{2kT}\right)v^2 \tag{2.10}$$

This is called the Maxwellian distribution function *for the speed*. Note that the only difference between $\eta(v)$ and $\eta(\mathbf{v})$ is the multiplicative factor $4\pi v^2$ in Equation (2.10). This factor comes from the fact that the volume of the spherical shell in Figure 2.4 is $4\pi v^2\,dv$. Thus *the number of (micro)states* with speed v is proportional to $4\pi v^2$. From this point of view:

$\eta(v)\,dv =$ (the number of states with speed v)(probable number of molecules in a particular state with speed v)

$$= (4\pi v^2)\left[N\left(\frac{m}{2\pi kT}\right)^{3/2}\exp\left(-\frac{mv^2}{2kT}\right)dv\right] \tag{2.11}$$

If we plot these different factors, we get something like Figure 2.6. This figure is analogous to Figure 3 in Chapter 3. In the present case, the *degeneracy* of the microstates with the speed v is $4\pi v^2$. Because this degeneracy increases with increasing values of v, the **most probable speed** of a molecule v_m is not zero (even though the most probable *velocity* is zero). To determine the most probable speed of a molecule we must solve the equation

$$\frac{1}{N}\left[\frac{d\eta(v)}{dv}\right]_{v=v_m} = 0$$

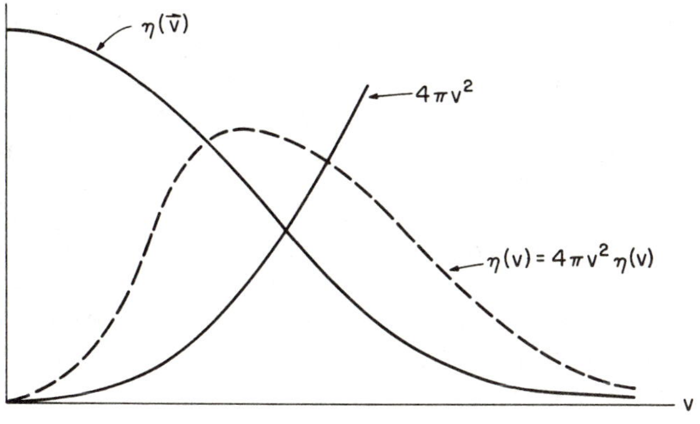

Figure 2.6

[the function $(1/N)\eta(v)$ is the distribution function for the speed of a *particular* molecule]. Using Equation (2.10), we find that (see problems)

$$\boxed{v_m = \left(\frac{2kT}{m}\right)^{1/2}} \tag{2.12}$$

Therefore the most probable speed increases if the temperature is increased. This fact is illustrated in Figure 2.7, where $\eta(v)/N$ is plotted for two values of the temperature.

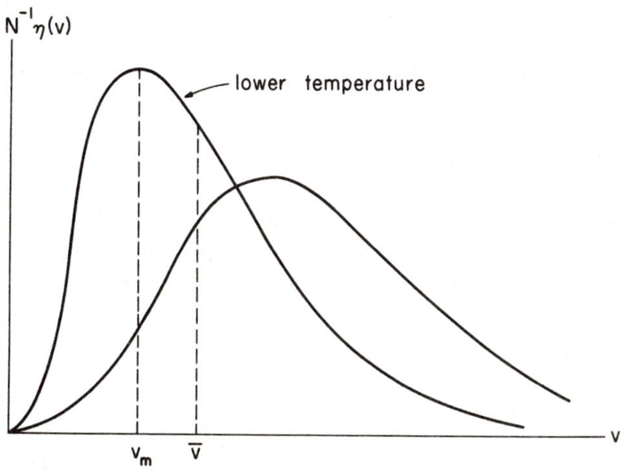

Figure 2.7

It will be recalled that the average *velocity* of the molecules is zero (because it is equally probable to find them traveling in any direction). The average *speed* of a molecule, on the other hand, is *not* zero. This is obvious when you remember that the speed is always a positive quantity, and hence cannot average to zero. To determine the average speed of a molecule $\bar{v}$ we use

$$\bar{v} = \frac{1}{N}\int_0^\infty v\eta(v)\,dv = \frac{4}{\sqrt{\pi}}\left(\frac{m}{2kT}\right)^{3/2}\int_0^\infty v^3 \exp\left(-\frac{mv^2}{2kT}\right)dv$$

(we again divide by N because we are interested in the average speed of a *single* molecule). This integral is the Gaussian integral G_3 (with $\beta = m/2kT$) evaluated in Appendix A. According to Equation (A.2) its value is $1/(2\beta^2)$. Hence we find the average speed is given by

$$\boxed{\bar{v} = \left(\frac{8kT}{\pi m}\right)^{1/2} = \frac{1}{\sqrt{\pi}}\,v_m} \tag{2.13}$$

We see that $\bar{v}/v_m \simeq 1.13$, and this is illustrated in Figure 2.7.

To illustrate the order of magnitude of some of these results, let us consider a perfect gas composed of O_2 molecules. Assume that the pressure is 1.013×10^5 newtons/m^2 (1 atmosphere), $T = 300°K$, and $V = 10^2$ cm^3. First of all, how many molecules are there in the system? For a perfect gas we have $pV = NkT$, and since $k = 1.38 \times 10^{-23}$ joule/deg, we obtain

$$N = \frac{1.013 \times 10^5 \times 10^{-4}}{1.38 \times 10^{-23} \times 300} = 2.45 \times 10^{21}$$

To appreciate the size of this number, one might compare it with the earth's population, which is roughly 3.5×10^9. Thus N is about 10^{12} times larger! To obtain the average speed of an O_2 molecule, we must first determine its mass. The molecular weight of an O_2 molecule is 32, and the value of one atomic mass unit is 1.66×10^{-27} kgm. Hence

$$m = 32 \times 1.66 \times 10^{-27} = 5.31 \times 10^{-26}\ \text{kgm}$$

Using Equation (2.13), we obtain for the average speed

$$\bar{v} = \left(\frac{8kT}{\pi m}\right)^{1/2} = \left(\frac{8 \times 1.38 \times 10^{-23} \times 300}{3.14 \times 5.31 \times 10^{-26}}\right)^{1/2}$$
$$= 445\ \text{m/sec}$$

This is quite fast, at least compared with ordinary experience (it is roughly 1,000 miles/hour). From these values we can begin to appreciate that there is a fantastic amount of activity going on in a little bit of gas.

Let us now estimate the probable number of these O_2 molecules that have a speed in the range $10^{-6}v_m$ of the value $v = v_m$. Since in this range of v

the distribution function $\eta(v)$ is essentially constant, the desired number is approximately $\eta(v = v_m)\,dv$, where $dv = 10^{-6}v_m$. Therefore we obtain

$$N\sqrt{\frac{2}{\pi}}\left(\frac{m}{kT}\right)^{3/2} v_m^2 \exp\left(-\frac{mv_m^2}{2kT}\right) \times 10^{-6}\, v_m$$

Using the fact that $v_m = (2kT/m)^{1/2}$, we find that this reduces to

$$\frac{4}{\sqrt{\pi}} Ne^{-1} \times 10^{-6} = 0.83 \times 10^{-6}N \simeq 2.04 \times 10^{15}$$

We see that even in this very small range of speeds ($10^{-6}\, v_m \simeq 0.04$ cm/sec) there is a very large number of molecules.

If one is interested in the probable number of molecules with a velocity component v_x larger than some value v_0, this is given by

$$\int_{v_0}^{\infty} \eta(v_x)\, dv_x = N\left(\frac{m}{2\pi kT}\right)^{1/2} \int_{v_0}^{\infty} \exp\left(-\frac{mv_x^2}{2kT}\right) dv_x \qquad (2.14)$$

Integrals of this type are essentially incomplete Gaussian integrals (since v_0 is not necessarily equal to zero). The best we can do in this case is to use tabulated values for an incomplete Gaussian integral of this type. The standard integral of this type is called the **error function** erf (x), defined by

$$\operatorname{erf}(x) = \frac{2}{\sqrt{\pi}} \int_0^x e^{-u^2}\, du \qquad (2.15)$$

Because of the factor $2/\sqrt{\pi}$, the error function goes from zero to unity as x goes from zero to infinity. The intermediate values are given in Appendix B. To evaluate an integral of the form (2.14), it is useful to first set $mv_x^2/2kT = u^2$ [so $dv_x = (2kT/m)^{1/2}\, du$]. Then (2.14) becomes

$$\frac{N}{\sqrt{\pi}} \int_{u_0}^{\infty} e^{-u^2}\, du \qquad \left[\text{where } u_0 = \left(\frac{m}{2kT}\right)^{1/2} v_0\right]$$

which is still not in the form of an error function. We can remedy this by writing it in the form

$$\frac{N}{\sqrt{\pi}} \int_0^{\infty} e^{-u}\, du - \frac{N}{\sqrt{\pi}} \int_0^{u_0} e^{-u^2}\, du = \tfrac{1}{2}N - \tfrac{1}{2}N \operatorname{erf}(u_0)$$

which now can be determined from the table for erf (x). For example, if we consider the oxygen gas described above and take $v_0 = 10^3$ m/sec, then $m/2kT = 6.45 \times 10^{-6}$ sec^2/m^2. Thus the probable number of molecules with $v_x \geq 10^3$ m/sec is given by

$$\tfrac{1}{2}N[1 - \operatorname{erf}(2.54)] = \tfrac{1}{2}N(1 - 0.99966) \simeq 4.1 \times 10^{17}$$

We shall now apply the results of this section to several specific problems.

PROBLEMS

1. Determine the number of molecules in 1 cm^3 of a perfect gas, when $T = 300°K$ and $p = 10^{-10}$ mm of mercury (a very good vacuum). One atmosphere = 760 mm of mercury.

2. (a) Derive Equation (2.12) for the most probable speed of a molecule.
 (b) What is the most probable value of the x component of the velocity, $(v_x)_m$ [or $(v_y)_m$, $(v_z)_m$]?
 (c) How do you reconcile the results of parts (a) and (b)?

3. A system consists of helium gas at standard pressure and temperature, in a volume of 10^3 cm^3. Assume that this is a perfect gas.
 (a) Determine the number of molecules in the system. How would this number change if the gas were N_2 (under the same conditions)?
 (b) What is the most probable velocity of a He molecule?
 (c) What is the most probable speed of a He molecule? Of a N_2 molecule under the same conditions?

4. (a) Let $\eta(\epsilon)\,d\epsilon$ be the probable number of molecules with a translational energy in the range $d\epsilon$ of ϵ. Using the relationship $\epsilon = \frac{1}{2}mv^2$, and the Maxwellian distribution function for the speed $\eta(v)$ determine $\eta(\epsilon)$.
 (b) Using the result of (a), determine the most probable energy ϵ_m of *a* molecule. Does it equal $\frac{1}{2}mv_m^2$?
 (c) What is the relative probability (ratio of probabilities) that a molecule will have a translational energy in the range $d\epsilon$ of $\epsilon = \epsilon_m$ compared with the range $d\epsilon$ of $\epsilon = 2\epsilon_m$? Evaluate this for $T = 200°K$ and $300°K$.

5. The root-mean-square speed of a molecule is defined as $v_{rms} = (\overline{v^2})^{1/2}$, where

$$\overline{v^2} = \frac{1}{N}\int_0^\infty v^2\,\eta(v)\,dv$$

 Determine v_{rms} for a perfect gas in equilibrium, and show how it is related to $\bar{v}$.

6. What is the average speed of those molecules with speeds above v_m?

7. What fraction of the molecules of a perfect gas have a speed (a) $v \geq v_m$, (b) $v \geq 2v_m$?

8. What fraction of the molecules of a perfect gas have $v_m \geq v_x \geq -v_m$?

9. What fraction of O_2 molecules have speeds greater than 5×10^4 cm/sec when (a) $T = 200°K$, (b) $T = 1{,}000°K$?

3. *APPLICATIONS OF THE MAXWELLIAN DISTRIBUTIONS*

The Maxwellian distribution functions arise in a variety of problems that involve low-density gaseous systems (whenever the intermolecular forces are negligible compared with the kinetic energy of the molecules). A few examples of such problems will now be discussed.

Consider a perfect gas that is in thermal equilibrium inside a container. Let us determine the probable number of molecules that will strike an area A of the container wall in a time dt. Since, in equilibrium, there is no preferred direction in space, we can take this element to be in the y,z plane, with the gas to the left of the element (see Figure 3.1). Now consider molecules with a velocity near some value $\mathbf{v}$. In a time dt they will move a distance

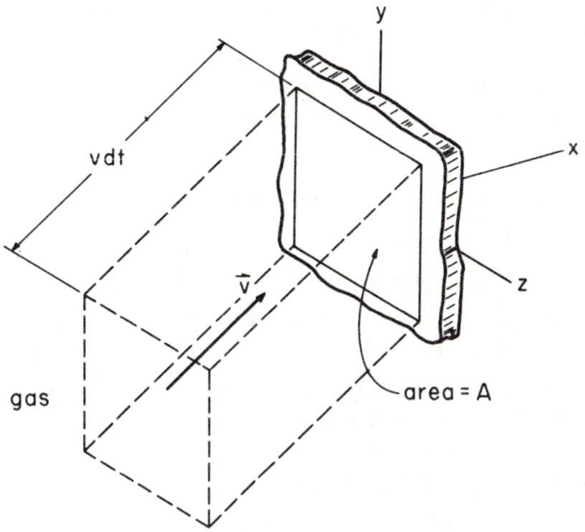

Figure 3.1

$v\,dt$. Therefore all these molecules that lie inside the volume shown in Figure 3.1 will strike A in a time dt, provided that $v_x \geq 0$ (if $v_x \leq 0$ the molecules are moving away from the wall). The volume of this spatial region equals the base A times the *normal* height $v_x\,dt$, or $\Omega = Av_x\,dt$. Now, according to Equation (2.2), the probable number of molecules in the volume Ω with a velocity in the range d^3v of v is given by

$$\int_\Omega d^3r\,\eta(\mathbf{r},\mathbf{v})\,d^3v = \left(\frac{\Omega}{V}\right)\eta(\mathbf{v})\,d^3v$$

where V is the volume of the gas [remember that $\eta(\mathbf{v}) = V\eta(\mathbf{r},\mathbf{v})$]. Using $\Omega = Av_x\,dt$, we obtain

The probable number of molecules with a velocity in the range d^3v of $\mathbf{v}$ ($v_x \geq 0$) that strike an area A in a time dt equals $v_x\eta(\mathbf{v})$ $d^3v(A/V)\,dt$. (3.1)

To obtain the *total* probable number of molecules that strike A in a time dt, we must simply integrate $v_x\eta(\mathbf{v})\,d^3v(A/V)\,dt$ over all velocities such that $v_x \geq 0$, or

$$\int_0^\infty dv_x \int_{-\infty}^\infty dv_y \int_{-\infty}^\infty dv_z\,v_x\eta(\mathbf{v})\left(\frac{A}{V}\right)dt$$

Substituting the Maxwellian distribution function for $\eta(\mathbf{v})$, Equation (2.4), we obtain

$$\left(\frac{N}{V}\right) A\, dt \left(\frac{m}{2\pi kT}\right)^{3/2} \int_0^\infty dv_x \int_{-\infty}^{\infty} dv_y \int_{-\infty}^{\infty} dv_z\, v_x \exp\left(-\frac{m\mathbf{v}^2}{2kT}\right)$$

The integral over v_x is the Gaussian integral G_1, and the other two integrals are each $2G_0$, so we obtain

$$\left(\frac{N}{V}\right) A\, dt \left(\frac{m}{2\pi kT}\right)^{3/2} \left(\frac{kT}{m}\right)\left(\frac{2\pi kT}{m}\right) = \left(\frac{N}{V}\right)\left(\frac{kT}{2\pi m}\right)^{1/2} A\, dt$$

If we set $n = N/V$ (the number density of the molecules), and use the fact that the average speed is $\bar{v} = (8kT/\pi m)^{1/2}$, we can write the final result in the form

The probable number of molecules that strike the wall per unit area and unit time $= \frac{1}{4}n\bar{v}$. (3.2)

This result is useful, for example, in estimating the rate at which gas will escape from a container that has a small pinhole in the wall (this process is called **molecular effusion**). In this case all molecules striking the pinhole will escape, instead of being reflected back into the interior. Such an estimate of the rate of effusion assumes, of course, that this rate is sufficiently slow for the remaining gas to be able to stay in thermal equilibrium [otherwise $\eta(\mathbf{v})$ would not be Maxwellian]. If the pinhole area is small enough, this is a good approximation.

Example

For the oxygen gas described in Section 2 we had $N = 2.45 \times 10^{21}$, $V = 10^2$ cm^3, and $\bar{v} = 445$ m/sec $= 4.45 \times 10^4$ cm/sec. In this case, the probable number of molecules that would escape through a pinhole with an area of 10^{-8} cm^2 per second would be

$$\frac{1}{4}\left(\frac{2.45 \times 10^{21}}{10^2}\right) \times 4.45 \times 10^4 \times 10^{-8} = 2.7 \times 10^{15}$$

which is about one out of every million molecules.

If N is the number of molecules in the container of volume V, then Equation (3.2) gives the rate of change of N during molecular effusion, namely,

$$\frac{dN}{dt} = -\left(\frac{N}{V}\right)\left(\frac{kT}{2\pi m}\right)^{1/2} A \tag{3.3}$$

It will be noted that dN/dt is proportional to $m^{-1/2}$, so that molecules of larger mass escape more slowly. In fact, at one time, measurements of the rates of effusion were used to determine the molecular weights of gases. A more recent application of Equation (3.3) involves the problem of separating U^{235} from U^{238}. The uranium, when combined with fluorine, forms the gas UF_6 (uranium hexafluoride). When this gas effuses through a small hole, the molecules of UF_6 that contain the lighter isotope U^{235} effuse faster than the others. In this way (by repeating the process several thousand times!) it is possible to obtain a fairly high degree of separation of the isotopes — thereby obtaining enriched U^{235}.

Another point to note about Equation (3.1) is that very few molecules with small values of v_x strike the wall in a time dt (because of the multiplicative factor v_x). Therefore the molecules that effuse through a hole are preferentially made up of those with the larger values of v_x. As a result the average translational energy per effused molecule is greater than the average translational energy $\frac{3}{2}kT$ of a molecule in the gas (see problems).

Another problem, which is quite similar, is to determine the *average force* that the wall must exert on the molecules in order to keep them confined. According to Newton's law, this average force must equal the change in the momentum ($m\mathbf{v}$) of all molecules that strike the wall per second. Therefore we want to estimate the total momentum change of all the molecules striking the area A in a time dt (Figure 3.1). For simplicity, assume that when a molecule strikes the wall it is specularly reflected, so that v_x just changes sign. In this case the change in its momentum is $-mv_x - (+mv_x) = -2mv_x$, where $v_x \geq 0$ is the x component of the velocity prior to collision with the wall. Now, according to Equation (3.1), the probable number of molecules with a velocity in the range d^3v of $\mathbf{v}$ that strike A in a time dt is $v_x\eta(\mathbf{v})\, d^3v\, (A/V)\, dt$. Multiplying this by the momentum change $-2mv_x$ of a molecule, and integrating over all relevant velocities ($v_x \geq 0$), we obtain:

The net momentum change of the molecules striking A in a time dt

$$= \int_0^\infty dv_x \int_{-\infty}^\infty dv_y \int_{-\infty}^\infty dv_z\, (-2mv_x)v_x\eta(\mathbf{v})\left(\frac{A}{V}\right) dt$$

If we divide this by dt, it must equal the average force exerted by the element A on the gas, or

$$\begin{aligned}\bar{F}_x &= -2mA\left(\frac{N}{V}\right)\left(\frac{m}{2\pi kT}\right)^{3/2} \int_0^\infty dv_x\, v_x^2 e^{-\beta v_x^2} \int_{-\infty}^\infty dv_y\, e^{-\beta v_y^2} \int_{-\infty}^\infty dv_z\, e^{-\beta v_z^2} \\ &= -2mA\left(\frac{N}{V}\right)\left(\frac{m}{2\pi kT}\right)^{3/2}\left[\frac{\sqrt{\pi}}{4}\left(\frac{2kT}{m}\right)^{3/2}\right]\left(\frac{2\pi kT}{m}\right)^{1/2}\left(\frac{2\pi kT}{m}\right)^{1/2} \\ &= -\left(\frac{NkT}{V}\right)A\end{aligned}$$

The force is negative since it must be directed toward the left in Figure 3.1. The magnitude of the force, divided by the area, is what we call the **pressure.** Hence

$$p \equiv \frac{|\bar{F}|}{A} = \frac{NkT}{V} \tag{3.4}$$

This is simply the *equation of state* for a perfect gas.

This derivation of the equation of state is *only* valid for a perfect gas. Furthermore, this method for obtaining the pressure would not work for other systems (solids, liquids, or imperfect gases) and hence it is of very limited use. In fact we have already seen how the equation of state can be determined for *any system*, using the partition function and the equation $p = kT(\partial \ln Z/\partial V)$. Using this method we derived Equation (3.4) in a much simpler fashion. The only advantage of the present derivation (which is frequently referred to as a "kinetic-theory" derivation) is that, in the particular case of a perfect gas, it shows very clearly how the pressure results from the collision of gas atoms with the wall.

PROBLEMS

1. Two containers of volume V_1 and V_2 are connected by a very small hole. Assume that the temperature of the gas in each container is maintained in equilibrium at temperatures T_1 and T_2, respectively. Determine the ratio of the pressures p_1/p_2 if the number of molecules in each container is constant in time.
2. In dry air the percentages of N_2, O_2, and Ar (argon) are respectively 78, 21, and 1. Determine their percentages in the air that has effused through a very small hole into a vacuum.
3. Using Equation (3.1), determine the rate at which the translational energy is transported through a small hole of area A. Then, using the result (3.2), determine the average translational energy per escaping molecule. Compare this with the average translational energy $3kT/2$ of a molecule in the original gas.
4. (a) Using Equation (3.1), determine the probability that an effused molecule will have a velocity component v_x in the range dv_x of v_x for *all* values of v_x. Denote this probability by $\hat{f}(v_x)\, dv_x$.

   (b) Determine the most probable value of v_x for an effused molecule, and compare this with the equilibrium value.

   (c) Compare the Maxwellian distribution $N^{-1}\,\eta(v_x)$ with $\hat{f}(v_x)$. Make a sketch of both distributions in one figure, indicating carefully their relative maximum values.
5. On the basis of the derivation of the equation of state given in this section, how would you expect the pressure to differ (i.e., for the same N, V, and T) if the molecules attract each other? Discuss what happens as a molecule approaches a wall, and how this affects its momentum exchange with the wall.
6. An interesting aspect of effusion is the fact that molecules with a large positive velocity in the x direction are preferentially "sifted out" of the equilibrium distribution. This suggests that the uncertainty of the velocity v_x should be less for the effused group than it is for the original equilibrium group of molecules. To investigate this point, we first obtain dimensionless distribution functions by introducing the dimensionless variable $u = (m/2kT)^{1/2}\, v_x$. Using the definition for the new distribution function, $g(u)\, du = f(v_x)\, dv_x$, the equilibrium distribution function is $g(u) = \pi^{-1/2} e^{-u^2}$.

(a) From problem (4a) obtain the corresponding $\hat{g}(u)$ for the effused molecules.

(b) Using the definition for the uncertainty, $h = -\int g \ln(g)\, du$ and $\hat{h} = -\int \hat{g} \ln(\hat{g})\, du$, determine the uncertainty in u for a molecule in each of these two groups (Note: $\int_0^\infty u \ln(u)\, e^{-u^2}\, du = -\gamma/4$, where $\gamma = 0.5772\ldots$ is Euler's constant).

(c) The total uncertainty in u for the effused molecules is $N_e\hat{h}$, where N_e is the number of effused molecules. The total uncertainty in u for the entire system is $(N - N_e)h + N_e\hat{h} = H(u)$, where N is the total number of molecules. Does $H(u)$ increase or decrease as more molecules effuse? Now a famous result of the second law of thermodynamics is that the entropy of an isolated system (like the present system) cannot decrease. Indicate several reasons why the present results, together with Equation (40) of Chapter 3, do not represent a violation of this result of the second law.

4. *PERFECT GAS IN AN EXTERNAL FIELD*

Now let us consider the effects of an external field acting on a perfect gas. The ever-present gravitational field is one example of such an external field. Another possibility is an external electric field acting on charged molecules (also see the sections on dielectric and paramagnetic systems). In any case, the basic assumption is that the gas is sufficiently rarefied for the interaction energy (between molecules) to be negligible compared with their individual kinetic energies. Also, for simplicity, we shall ignore the internal motions of the molecules and treat them as point particles.

Under these conditions we have seen (Section 4 of Chapter 3) that the probability that a particle is in the region d^3r of $\mathbf{r}$ with a velocity in the range d^3v of $\mathbf{v}$ is given by

$$f(\epsilon)\, d^3r\, d^3v, \qquad \text{where } f(\epsilon) = \mathfrak{z}^{-1}\, e^{-\beta\epsilon} \tag{4.1}$$

The energy of the particle ϵ is

$$\epsilon = \tfrac{1}{2}m\mathbf{v}^2 + \Phi(\mathbf{r}) \tag{4.2}$$

where $\Phi(\mathbf{r})$ is its potential energy due to the external field. Thus the situation is very similar to that of the perfect gas in Section 2, except that now the potential energy $\Phi(\mathbf{r})$ makes the probability (4.1) depend on the position $\mathbf{r}$ of the particle.

To be specific, let us consider the effects produced by a gravitational field. If z is the height above the earth's surface, then the potential energy

$\Phi(\mathbf{r})$ can be approximated over small distances by $\Phi(\mathbf{r}) = mgz$ (Chapter 2, Section 2), where $g = 980$ cm/sec². Then (4.2) becomes

$$\epsilon = \tfrac{1}{2}m\mathbf{v}^2 + mgz \tag{4.3}$$

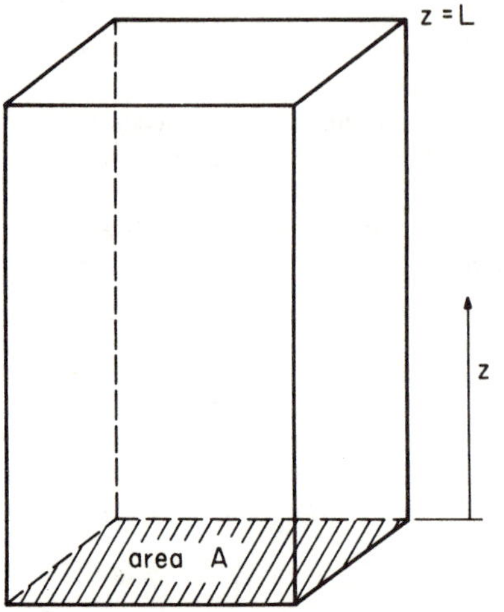

Figure 4.1

If we consider an enclosed column of gas, shown in Figure 4.1, of height L and base area A, the molecular partition function is then

$$\mathfrak{z} = \int_A dx\, dy \int_0^L dz \int_{-\infty}^{\infty} d^3v\, e^{-\beta\epsilon}$$

where the first three integrals are the integrals over the volume $\int_V d^2r$. Substituting the expression (4.3) into $\mathfrak{z}$, we have

$$\mathfrak{z} = A \int_0^L e^{-\beta mgz}\, dz \int_{-\infty}^{\infty} d^3v\, e^{-\beta m\mathbf{v}^2/2}$$

or

$$\mathfrak{z} = \left(\frac{2\pi}{m\beta}\right)^{3/2} \frac{A}{\beta mg}(1 - e^{-\beta mgL}) \tag{4.4}$$

We note that if βmgL is small, then $e^{-\beta mgL} \simeq 1 - \beta mgL$, and (4.4) reduces to

$$\mathfrak{z} \simeq \left(\frac{2\pi}{m\beta}\right)^{3/2} AL = \left(\frac{2\pi}{m\beta}\right)^{3/2} V$$

Thus if L is small, the gravitational field has no effect.

Now let us return to (4.1). Using (4.3), we find that

$$\mathfrak{z}^{-1}e^{-\beta[(1/2)m\mathbf{v}^2+mgz]}d^3r\, d^3v = \text{probability that a particle is in the region } (d^3r, d^3v) \text{ about } (\mathbf{r}, \mathbf{v}) \tag{4.5}$$

where $\mathfrak{z}$ is given by (4.4). We see that this probability decreases as z increases — which means that the particle is more likely to be found near $z = 0$ than at large heights ($z \gg 0$). The probable number of atoms in the region d^3r of $\mathbf{r}$ is given by

$$n(\mathbf{r})\, d^3r = \int_{-\infty}^{\infty} d^3v\, Nf(\epsilon)\, d^3r = N\frac{\beta mg}{A}(1 - e^{-\beta mgL})^{-1}e^{-\beta mgz}\, d^3r \tag{4.6}$$

which clearly decreases as z increases. Because $n(\mathbf{r})$ varies with the elevation, "*the* pressure" of this system is not defined. That is to say, we can only speak of a "local" pressure. In the atmosphere (where L is infinite), we can *define* a "local pressure" by

$$p(\mathbf{r}) \equiv n(\mathbf{r})kT = \frac{Nmg}{A}e^{-\beta mgz} \tag{4.7}$$

which is sometimes called the *law of atmospheres*. The fact that "*the* pressure" is not defined can be seen from the fact that total partition function $Z = \mathfrak{z}^N$ is *not* simply a function of the volume $V = AL$, but of A and L separately. Therefore the usual relationship

$$p = kT \left(\frac{\partial \ln Z}{\partial V}\right)_T$$

has no meaning.

Although the pressure is not defined, the internal energy of this system is obtained in the usual fashion from

$$U = -N \frac{\partial}{\partial \beta} \ln \mathfrak{z}$$

If $L = \infty$ (the atmosphere), then Equation (4.4) yields

$$U = \tfrac{3}{2}NkT + NkT = \tfrac{5}{2}NkT \tag{4.8}$$

and we see that the gravitational potential energy has increased the internal energy by a factor of NkT (the factor $\frac{3}{2}\, NkT$ being due to the average translational energy of the particles).

An important application of Equation (4.6) is that it can be used to directly determine the value of Boltzmann's constant. This method was used by Jean B. Perrin and his co-workers in the early part of this century. They suspended colloidal particles of gamboze (a yellowish gum resin) in a liquid and observed the number of these particles at various heights z after the particles obtained an equilibrium distribution. If we assume that these particles behave like a perfect gas, their number density should then be given by Equation (4.6). The effect of the liquid is to reduce the mass of the particles (by giving them a buoyancy force). If ρ is the mass density of gamboze, and v_0 is the volume of a colloidal particle, then the effective mass of a colloidal particle should be

$$m = v_0(\rho - \rho_{\text{liquid}}) \tag{4.9}$$

Using (4.6), we then find that (see Figure 4.2)

$$\frac{\text{probable number in } dz \text{ about } z_1}{\text{probable number in } dz \text{ about } z_2} = e^{-\beta mg(z_1 - z_2)} \tag{4.10}$$

By making a large number of observations (to obtain the average number of particles in the regions about z_1 and z_2), using a microscope with a very limited field of vision, Perrin obtained several hundred observations at each of two levels z. The fixed volume of observation, determined by the field of vision and depth of the focal plane of the microscope, was small enough to make the number of visible particles small (less than five or six), so that they could be readily counted. In this way he obtained the *average* number of particles in a fixed volume about z_1 and z_2. The ratio of these average

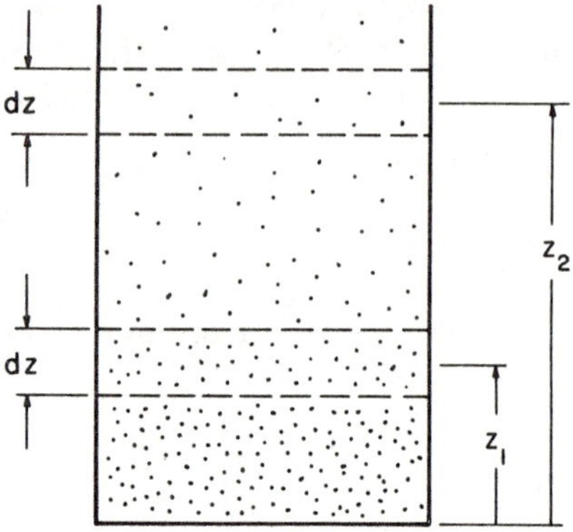

Figure 4.2

numbers then gives an experimental value for the left side of Equation (4.10). Knowing z_1, z_2, m, g, and T, he could then determine the value of the Boltzmann constant k from the right side of (4.10) (an example is given in the problems). Using the experimental value of the gas constant R then yields a value for Avogadro's number $N_0 = R/k$. He obtained values of R between 6 and 7 $\times 10^{23}$ molecules/gm-mole. The best value by this method was $R = 6.09 \times 10^{23}$ which is remarkably close to the presently accepted value of 6.0248×10^{23}. It is all the more remarkable when one considers that, by observing less than six particles at a time, it is possible to accurately predict a number of the order of 10^{23}!†

The rather weak gravitational field can be replaced by stronger forces, which then produce correspondingly larger effects. One method involves the use of an ultracentrifuge, which spins a colloidal suspension about an axis with an angular velocity ω. The outward radial force on a colloidal particle (the centrifugal force) is then $m\omega^2 r$, where r is the distance from the axis of rotation and m is again given by (4.9). For large values of ω one can readily see that large density variations will occur over small ranges of r. In fact if ω is too large, the colloidal particles will entirely displace the liquid, and the two substances will separate into distinct regions. In this case, of course, the colloidal particles do not behave like an ideal gas, and the present analysis is not applicable.

PROBLEMS

1. In Figure 4.1 consider an imaginary horizontal slab of gas, the bottom of which is at $z = h$ and the top at $z = h + dh$. According to Equation (4.7) the pressure of the gas outside this slab is different at the top and the bottom. Why does the slab of gas remain at a fixed height? Prove that, in fact, (4.7) is necessary for this slab to remain at a fixed height.

†For a general review of the methods used to obtain Avogadro's number see D. L. Anderson, Resource Letter ECAN-1, *American Journal of Physics* **34,** 1 (1966). A more detailed discussion of Perrin's work can be found in S. Glasstone, *Textbook of Physical Chemistry* (Princeton, N.J.: D. Van Nostrand Co., Inc., 1940).

2. Assume that each component of the air has a density variation given by Equation (4.7). At the surface of the earth ($z = 0$) the percentages of N_2 and O_2 are approximately 78 and 21, respectively. At 300 km what is the relative concentration of N_2 to that of O_2, if $T = 260°K$?

3. The mean molecular weight of air is 29. What is the pressure at 100 km if $T = 240°K$, assuming that the law of atmospheres is valid? Compare this with the average atmospheric value of 4.2×10^{-4} mm Hg (at this altitude). The pressure at the ground is one atmosphere (760 mm Hg).

4. The actual gravitational potential energy of a molecule is $\Phi = -c/r$, where c is a positive constant and r the distance from the center of the earth ($r_0 \leq r \leq \infty$, r_0 being the earth's radius). Consider all of outer space to be the volume V of the gas. Using spherical spatial coordinates, evaluate the molecule partition function

$$\mathfrak{z} = \int_V d^3r \int_{-\infty}^{\infty} d^3v \; e^{-\beta\epsilon}$$

Can you explain the significance of this result?

5. The experiments of Perrin and his co-worker are described in S. Glasstone *Textbook of Physical Chemistry* (D. Van Nostrand Co., Inc., 1940), pp. 249–252. In one case they found a value of $10^{0.481}$ for the left side of Equation (4.10), when the temperature was 25°C and $z_2 - z_1$ was only 0.01 mm. The average radius of the colloidal particles was 0.368×10^{-4} cm, and the mass density was $\rho = 1.194$ gm/cm^3. The density of the liquid (water) at this temperature is 0.997 gm/cm^3. What value was obtained for Boltzmann's constant k?

6. A centrifuge consists of a cylinder of length L and cross-sectional area A (small). The cylinder is suspended from one end and rotates about an axis that is perpendicular to its length. Derive an expression for the density $n(r)$ of the colloidal particles as a function of the distance r from the axis of rotation. Obtain the normalized form for $n(r)$ if there are N particles in the cylinder. Assume that their density is sufficiently small for them to be treated as a perfect gas.

5. *THE IMPERFECT GAS*

As the density of a gas is increased, the forces between the molecules become more important. As a result of these intermolecular forces, the perfect-gas equation of state

$$pV = NkT \tag{5.1}$$

no longer accurately describes the equation of state of the gas. In this section we shall obtain the first correction to Equation (5.1), which is produced by the intermolecular forces. To obtain this correction we shall have to obtain an approximate evaluation of the partition function in the case when the interaction energy is not zero.

In Chapter 2 we discussed the fact that the energy of a system of N molecules often can be written in the form

$$E = \sum_{i=1}^{N} \epsilon_i + \sum_{i>j}^{N} \Phi(r_{ij}) \tag{5.2}$$

The first summation in (5.2) represents the energy of the molecules when there are no forces between the molecules. Only this term is present in the case of a perfect gas. The second summation in (5.2) represents the *interaction potential energy* between the *molecules* (not between the atoms in one molecule, which is contained in the ϵ's). Thus $\Phi(r_{ij})$ is the potential energy between the molecules i and j, which are separated by a distance $r_{ij} = |\mathbf{r}_i - \mathbf{r}_j|$. The summation is for $i > j$ so that the interaction energy is not counted twice $\left[\text{it could also be written } \frac{1}{2}\sum_{i \neq j}^{N} \Phi(r_{ij})\right]$.

Now the partition function can be written

$$Z = \int_V d^3r_1 \cdots d^3r_N \int d(\text{remaining variables})\, e^{-\beta E} \tag{5.3}$$

where the first group of integrals are over the volume of the system, for the position of the center of mass $\mathbf{r}_i$ of each molecule. The second group of integrals in (5.3) contain the translational and "internal" variables (describing the rotational and vibrational motion). If we substitute (5.2) into (5.3), then we obtain

$$Z = \int_V d^3r_1 \cdots d^3r_N \, e^{-\beta\Sigma\Phi(r_{ij})} \int e^{-\beta\Sigma\epsilon_i}\, d(\text{remaining variables}) \tag{5.4}$$

Now we note that the pressure is given by

$$p = kT\left(\frac{\partial \ln Z}{\partial V}\right) \tag{5.5}$$

and since only the first group of integrals in (5.4) depend on V, we have

$$\ln Z = \ln\left(\int_V d^3r_1 \cdots d^3r_N \, e^{-\beta\Sigma\Phi(r_{ij})}\right) + (\text{terms independent of } V)$$

If we substitute this in (5.5), we see that

$$p = kT\frac{\partial(\ln Z_{\text{conf}})}{\partial V} \tag{5.6}$$

where

$$Z_{\text{conf}} = \int_V d^3r_1 \cdots d^3r_N \, e^{-\beta\Sigma\Phi(r_{ij})} \tag{5.7}$$

is referred to as the **configurational** part of the **partition function.** If $\Phi(r_{ij}) = 0$ (no intermolecular forces), then $Z_{\text{conf}} = V^N$ and (5.6) immediately yields the perfect-gas equation of state (5.1). The whole problem, therefore, is how to evaluate (5.7) when $\Phi(r)$ is not zero. This is a complicated integral, because it involves many variables and the function $\Phi(r)$ is not simple (more on this later), so we must be contented with an approximate evaluation of (5.7).

To simplify matters we can note that when a gas is not too dense, most of the interactions between the molecules are due to interaction between *pairs* of molecules (and not among three or four molecules at a time). To make use of this observation let us rewrite the integrand of Z_{conf} in the following form:

$$\exp\left(-\beta\sum_{i>j}\Phi(r_{ij})\right) = \prod_{i>j} e^{-\beta\Phi(r_{ij})} = \prod_{i>j}\left[1 + \left(e^{-\beta\Phi(r_{ij})} - 1\right)\right] \tag{5.8}$$

where Π represents a *product.* Thus

$$\begin{aligned}\prod_{i>j} e^{-\beta\Phi(r_{ij})} &= e^{-\beta\Phi(r_{21})} \times e^{-\beta\Phi(r_{31})} \times e^{-\beta\Phi(r_{32})} \times \cdots \\ &= e^{-\beta[\Phi(r_{21})+\Phi(r_{31})+\Phi(r_{32})+\ldots]}\end{aligned}$$

which is the same as the left side of (5.8). The last expression in Equation (5.8) is obtained simply by adding and subtracting one in each term in the product. This looks like a strange thing to do — but note that the terms $(e^{-\beta\Phi(r_{ij})} - 1)$ are zero if $\Phi(r_{ij}) = 0$. Therefore the quantities

$$u_{ij} \equiv e^{-\beta\Phi(r_{ij})} - 1 \tag{5.9}$$

represent the deviation from a perfect gas due to the interaction between *pairs* of molecules. Using the notation (5.9), the integrand of Z_{conf}, Equation (5.8), can be written

$$\begin{aligned}\prod_{i>j}(1 + u_{ij}) &= (1 + u_{21})(1 + u_{31})(1 + u_{32})\cdots \\ &= 1 + (u_{21} + u_{31} + u_{32} + \cdots) \\ &\qquad + \text{terms containing higher powers of } u\end{aligned}$$

The terms $u_{21} + u_{31} + \cdots = \sum_{i>j} u_{ij}$ represent the interactions between *pairs* of molecules, whereas the higher powers of u (such as $u_{21}\,u_{31}$) vanish unless three or more molecules are in interaction. If we keep just the terms that represent the interaction between one pair of molecules at a time, then

$$Z_{\text{conf}} \simeq \int_V d^3r_1\cdots d^3r_N\left(1 + \sum_{i>j}^{N} u_{ij}\right) = V^N + \int_V d^3r_1\cdots d^3r_N \sum_{i>j}^{N} u_{ij}$$

The last integral can also be written as the sum of integrals

$$\sum_{i>j}^{N}\int d^3r_1\cdots d^3r_N\, u_{ij} = V^{N-2}\sum_{i>j}^{N}\int_V d^3r_i\, d^3r_j\, u_{ij}$$

Each integral in this summation has the same value [they differ only by the labels (i, j) on the integration variables]. Moreover, there are $\frac{1}{2}N(N-1)$ terms in this sum [i can be any of N numbers, and j any of the $(N-1)$ remaining numbers; but then divide by 2 because $i > j$]. Therefore

$$Z_{\text{conf}} \simeq V^N + V^{N-2}\,\tfrac{1}{2}N(N-1)\int_V d^3r_1\, d^3r_2\, u_{12}(|\mathbf{r}_1 - \mathbf{r}_2|)$$

We can simplify the remaining integrals by introducing the variable

$$\mathbf{r} = \mathbf{r}_1 - \mathbf{r}_2 \qquad (\text{or set } \mathbf{r}_2 = \mathbf{r} + \mathbf{r}_1)$$

Then, since $d^3r_1\, d^3r_2 = d^3r_1\, d^3r$, we have

$$Z_{\text{conf}} \simeq V^N + V^{N-2}\,\tfrac{1}{2}N(N-1)\int_V d^3r_1 \int_V d^3r\, u(r)$$

The first integral gives another factor of V, and for large N we also can set $N(N-1) \simeq N^2$, so we finally obtain

$$Z_{\text{conf}} \simeq V^N - V^{N-1}N^2B(T) \tag{5.10}$$

where

$$\boxed{B(T) = \tfrac{1}{2}\int d^3r\,(1 - e^{-\beta\Phi(r)})} \tag{5.11}$$

The quantity $B(T)$ is known as the **second virial coefficient.** Note that since $\Phi(r)$ goes to zero if r is larger than a few angstroms (say 5×10^{-8} cm), the integrand in (5.11) goes to zero unless r is very small. Because of this $B(T)$ is independent of the volume of the system.

The expression (5.10) for Z_{conf} contains the *first correction* due to the intermolecular forces. Therefore, to the same accuracy, we can write

$$\ln Z_{\text{conf}} \simeq \ln V^N[1 - V^{-1}N^2B(T)] = \ln V^N + \ln[1 - V^{-1}N^2B(T)]$$

or, using $\ln(1-x) \simeq -x$ for small x, this reduces to

$$\ln Z_{\text{conf}} \simeq \ln V^N - V^{-1}N^2B(T) \tag{5.12}$$

We can now obtain the approximate equation of state by substituting (5.12) into (5.6):

$$p = kT\frac{\partial}{\partial V}[N\ln V - V^{-1}N^2B(T)] = kT\left[\frac{N}{V} + \frac{N^2}{V^2}B(T)\right]$$

or

$$\boxed{pV = NkT\left[1 + \frac{NB(T)}{V}\right]} \tag{5.13}$$

What we have obtained is the first correction in the so-called **virial series**

$$pV = NkT\left[1 + \frac{NB(T)}{V} + \frac{N^2C(T)}{V^2} + \cdots\right] \tag{5.14}$$

where $B(T)$, $C(T)$, and so on are called the second, third, ... virial coefficients. This expression is a power series in (N/V), which is useful only if (N/V) is not too large.

To estimate the second virial coefficient $B(T)$, we must use some expression for the interaction potential $\Phi(r)$. The nature of $\Phi(r)$ was discussed in Chapter 2 and found to be of the form shown by the solid curve in Figure 5.1. For small values of r, $\Phi(r)$ drops rapidly as r increases (corresponding to a strong repulsive force). For larger values of r, $\Phi(r)$ slowly increases (corresponding to a weak attractive force). An approximate form for $\Phi(r)$ might be the Lennard-Jones potential (see Chapter 2).

$$\Phi(r) = \epsilon_0\left[\left(\frac{r_0}{r}\right)^{12} - 2\left(\frac{r_0}{r}\right)^{6}\right] \tag{5.15}$$

Although it is possible to evaluate $B(T)$ using Equation (5.15), the analysis is somewhat cumbersome and involves numerical evaluations of an infinite series. To avoid such problems, we shall consider another approximate potential, given by

$$\Phi(r) = \begin{cases} \infty, & r < 2r_s \\ -\epsilon_s, & 2r_s < r < 2r_a \\ 0, & r > 2r_a \end{cases} \tag{5.16}$$

This is known as a **square-well potential** and is shown as the dotted curve in Figure 5.1. In this case the molecules behave like hard spheres (of radius

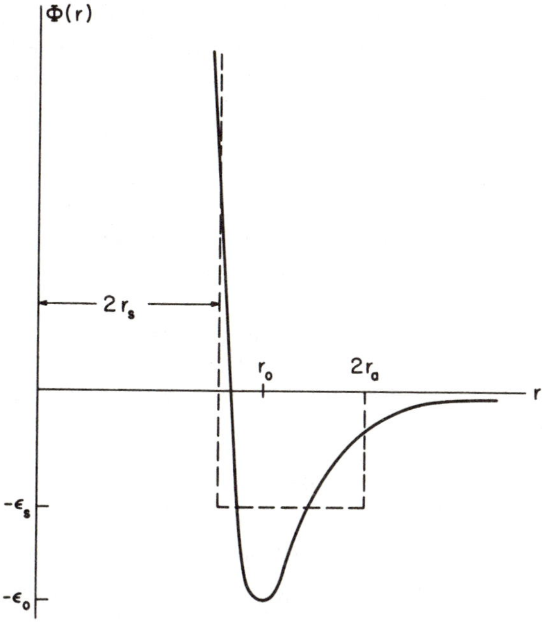

Figure 5.1

r_s) with an attractive (binding) energy ϵ_s for $2r_s < r < 2r_a$. Another type of approximate potential is illustrated in the problems.

If we introduce spherical coordinates in Equation (5.11) for $B(T)$, and use the potential (5.16), we obtain

$$B(T) = \tfrac{1}{2}\int_0^\infty dr\, r^2(1 - e^{-\beta\Phi(r)}) \int_0^\pi \sin\theta\, d\theta \int_0^{2\pi} d\phi$$
$$= 2\pi \int_0^{2r_s} r^2\, dr + 2\pi \int_{2r_s}^{2r_a} dr\, r^2(1 - e^{+\beta\epsilon_s})$$

where we have broken up the r integration into the three regions in (5.16). We then obtain

$$B(T) = \frac{16\pi}{3}\left[r_s^3 - (r_a^3 - r_s^3)(e^{\beta\epsilon_s} - 1)\right] \tag{5.17}$$

If we substitute this result into Equation (5.13)

$$pV = NkT\left[1 + \frac{NB(T)}{V}\right] \tag{5.13}$$

we obtain an approximate equation of state that contains the first correction due to intermolecular forces.

Notice that the temperature dependence of $B(T)$ in (5.17) is due entirely to the *attractive* interactions [for if $\epsilon_s = 0$, then $B(T)$ would not depend on T]. Since the attractive force between different types of molecules is quite different, so is the temperature dependence of $B(T)$. This is illustrated by the experimental curves shown in Figure 5.2 (also see problems). In Figure 5.2 the quantity $\tilde{B}(T)$ — which is what is usually tabulated — is related to $B(T)$ by

$$\tilde{B}(T) = N_0 B(T); \qquad N_0 = \text{Avogadro's number} \tag{5.18}$$

The expression (5.17) can be simplified further if $1 \gg \beta\epsilon_s$. In that case we can set $e^{\beta\epsilon_s} \simeq 1 + \beta\epsilon_s$, and then (5.17) reduces to

$$B(T) \simeq \frac{16\pi}{3}\left[r_s^3 - r_a^3\beta\epsilon_s\right] \tag{5.19}$$

where we have also used the approximation $r_s^3(1 + \beta\epsilon_s) \simeq r_s^3$. Now let us define two parameters a and b by

$$a = \frac{16\pi}{3}\,\epsilon_s r_a^3 N_0^2, \qquad b = \frac{16\pi}{3}\,r_s^3 N_0 \tag{5.20}$$

where N_0 is Avogadro's number. Then, from (5.18) and (5.19), we have

$$\tilde{B}(T) = b - \frac{a}{RT} \qquad (R = N_0 k) \tag{5.21}$$

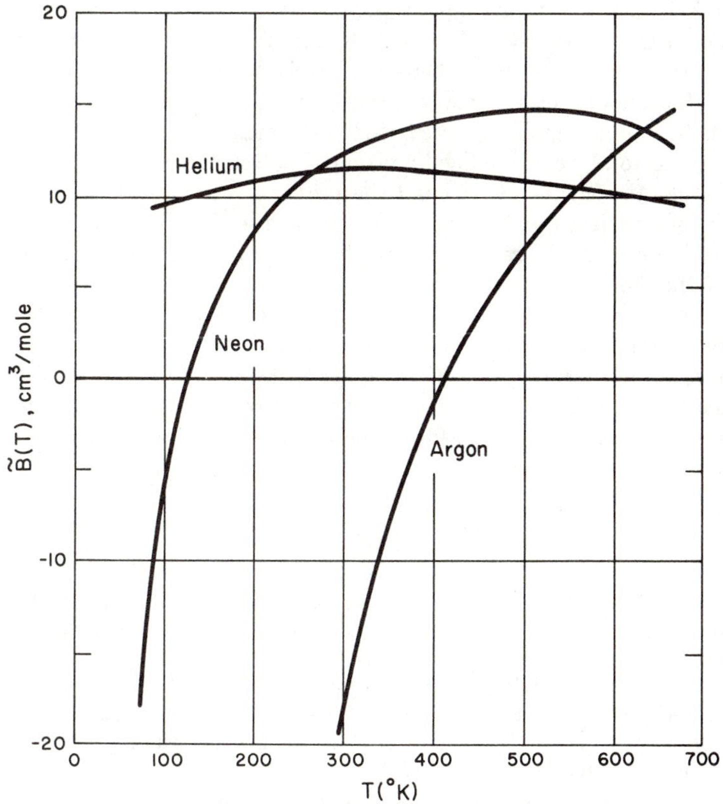

Figure 5.2

If we substitute this into (5.13), we obtain

$$pV = NkT\left[1 + \frac{NB(T)}{V}\right] = nRT\left[1 + \frac{n\tilde{B}(T)}{V}\right]$$
$$= nRT\left[1 + \frac{n}{V}\left(b - \frac{a}{RT}\right)\right]$$
$$\simeq nRT\left[\left(1 - \frac{nb}{V}\right)^{-1} - \frac{na}{VRT}\right]$$

where $n = N/N_0$ is the number of moles, and we have used

$$\left(1 - \frac{nb}{V}\right)^{-1} \simeq 1 + \frac{nb}{V}$$

A little rearranging then yields

$$\left[p + a\left(\frac{n}{V}\right)^2\right](V - nb) = nRT \qquad (5.22)$$

which is the **van der Waals equation of state.** We can see from (5.20) that the constant a in the van der Waals equation is related to the attractive force between the molecules, whereas b is related to the repulsive interactions. In fact b is the volume that one mole of atoms *cannot* occupy because of the infinite repulsive force between the molecules. The factor nb is therefore the total "excluded volume" in the system, and hence is subtracted from V. On the other hand the pressure is reduced from what it would be in a perfect gas, because of the factor $a(n/V)^2$. The fact that the attractive forces should reduce the pressure can be easily understood in terms of the momentum exchange of the atoms with the wall (Section 3). When an atom moves toward the wall, it is attracted back toward the interior by the attractive force of the other atoms. This causes it to slow down as it approaches the wall. Consequently it does not strike the wall as hard as it would if the attractive forces were not present (e.g., perfect gas). Therefore the pressure is reduced by the attractive interactions, and this leads to the correction term $a(n/V)^2$ in (5.22).

PROBLEMS

1. The value of ϵ_s in the square-well potential is usually taken to be about $0.58\epsilon_0$, where ϵ_0 is the Lennard-Jones parameter in Equation (5.15). From Table 2 in Chapter 2 we have

   $\epsilon_0 = 1.42 \times 10^{-15}$ (He), 4.83×10^{-15} (Ne), 16.6×10^{-15} (Ar) ergs

   Using these values and Equation (5.17), give a qualitative explanation of the fact that $B(T)$ is nearly independent of T for He, and constant for Ne at lower values of T than for Ar (see Figure 5.2).

2. The following experimental values of $\tilde{B}(T)$ have been obtained for neon (Figure 5.2)

T (°K)	373	273	235	173	123	100
$\tilde{B}(T)$ (cm³/mole)	13.5	11	9.5	6	0	−7.5

   Plot $\tilde{B}(T)$ vs. $(1/T)$. Obtain a straight-line approximation through the first *five* points and thereby estimate the van der Waals constants in (5.21). Using (5.17), explain why the experimental point at $T = 100$°K does *not* fall on this straight line (note $\epsilon_s \simeq 2.7 \times 10^{-15}$ ergs for Ne).

3. It is found that the square-well potential most closely reproduces the results obtained from the Lennard-Jones potential if

$$\epsilon_s = 0.58\epsilon_0 \qquad \text{and} \qquad r_s \simeq 0.41 r_0$$

   Using the values of (ϵ_0, r_0) for argon (Table 2, Chapter 2), and Equation (5.17), plot $\tilde{B}(T) = N_0\, B(T)$ for 700°K > T > 250°K. Use the fact that $\tilde{B}(T_B) = 0$, for $T_B = 412$°K (known as Boyle's temperature) in order to fix the value of $(r_a/r_s)^3$. Compare your results with the experimental curve in Figure 5.2.

[Warning: Three significant figures are required at the beginning to obtain $\tilde{B}(T)$ to two significant figures.]

4. Another approximate potential $\Phi(r)$ which can be used relatively easily, is the Sutherland potential

$$\Phi(r) = \begin{cases} \infty, & r < 2r^* \\ -\epsilon^* \left(\dfrac{2r^*}{r}\right)^6, & r > 2r^* \end{cases}$$

where ϵ^* and r^* are constants. This potential corresponds to a hard sphere of radius r^*, with an attractive potential that increases as r^{-6} (as in the Lennard-Jones potential). Evaluate $B(T)$ in the case where $\epsilon^*/kT \ll 1$ (so that $e^{\beta\epsilon^*} \simeq 1 + \beta\epsilon^*$ can be used in the integrand of Equation (5.11). Determine the relationship between the van der Waals parameters (a, b) and the constants (ϵ^*, r^*).

5. Using the Sutherland potential in Problem 4, obtain an *exact* expression for $B(T)$ by expanding the integrand of Equation (5.11) in an infinite series.

6. THE EQUIPARTITION LAW

In the previous sections we were concerned only with the translational motion of molecules in a perfect gas. Now we turn to more general considerations, which are applicable to any system. In this section we shall derive a famous (or infamous) prediction of classical statistical mechanics known as the equipartition of energy. We shall then illustrate how this prediction is modified by the quantization of energy.

We begin with the classical expression for the partition function:

$$Z = \int_V \cdots \int_{-\infty}^{\infty} e^{-\beta E(\mathbf{r}_1, \ldots, \mathbf{v}_N)}\, d^3r_1 \cdots d^3v_N \tag{6.1}$$

Assume that the energy of the system is the sum of two parts

$$E = cs^2 + E' \tag{6.2}$$

where s is one of the integration variables, c is a positive constant, and E' *does not contain the variable* s. For example s might be the v_x component of some atom, or the spatial variable z of some atom. The important point is that the variable s, whatever it is, should appear in $E(\mathbf{r}_1, \ldots, \mathbf{v}_N)$ as a *quadratic* term. This is always true of the velocity components v_x, v_y, and v_z, but it is not always true of the spatial variables x, y, and z of the various particles. One of the important cases where the spatial variables enter in this quadratic fashion is in the case of harmonic oscillators. We shall consider this case below.

Now if E is of the form (6.2), then Z can be written in the form

$$Z = \int_{-\infty}^{\infty} ds\, e^{-\beta cs^2} \int \cdots \int e^{-\beta E'}\, d(\text{remaining variables})$$

because E' does not depend on s. Here we are assuming that s can range from $-\infty$ to $+\infty$. Let us denote the last group of integrals by Z', so that

$$Z = Z' \int_{-\infty}^{\infty} ds\, e^{-\beta cs^2}$$

The integral involving s is the Gaussian integral $2G_0$, so we find that

$$Z = \left(\frac{\pi}{\beta c}\right)^{1/2} Z'$$

If we substitute this expression into the general expression for the internal energy $U = -\partial \ln Z/\partial\beta$, we obtain

$$U = \frac{1}{2}\frac{\partial}{\partial\beta} \ln\left(\frac{\beta c}{\pi}\right) - \frac{\partial}{\partial\beta} \ln Z' = \frac{1}{2\beta} - \frac{\partial}{\partial\beta} \ln Z = \frac{1}{2} kT + U'$$

where we have set $U' \equiv -\partial \ln Z'/\partial\beta$. We see that the internal energy is made up of the sum of two terms $\frac{1}{2}kT + U'$. The term U' is the contribution to the internal energy due to all the variables in Z except the variable s. The motion associated with the variable s contributes the amount $\frac{1}{2}kT$ to the internal energy — provided that s appears in the energy in a quadratic fashion [Equation (6.2)]. In a similar way, if there are other variables in Z' that appear in the energy in a quadratic form, then they will also contribute a term $\frac{1}{2}kT$ to the internal energy. This means that if the energy E is the sum of N quadratic terms (as it frequently is), then the internal energy of that system should be $N(\frac{1}{2}kT)$. This result is often described by saying that *the average energy associated with every quadratic variable in* $E(\mathbf{r}_1, \ldots, \mathbf{v}_N)$ is $\frac{1}{2}kT$. In other words the energy is, on the average, distributed equally over all these types of motion. The result above therefore is referred to as the **equipartition of energy.** It should be noted that the result above is independent of the magnitude of the constant c in Equation (6.2). Thus, for example, in the case of a perfect monatomic gas, where

$$E = \sum_{i=1}^{N} \tfrac{1}{2}m\mathbf{v}_i^2$$

the constant c corresponds to $\frac{1}{2}m$. In this case there are $3N$ quadratic terms, so classical statistical mechanics predicts that

$$U = 3N(\tfrac{1}{2}kT)$$

and this result is independent of the mass of the atoms.

To illustrate the limitation of the equipartition law that results from the quantization of the energy, let us consider a case that frequently arises in statistical mechanics. This is the case of a *harmonic oscillator.* In classical mechanics the energy of a harmonic oscillator is

$$\epsilon = \tfrac{1}{2}mv_x^2 + \tfrac{1}{2}kx^2$$

Since there are two quadratic variables, the equipartition law predicts that the *average energy of a harmonic oscillator is* $2(\tfrac{1}{2}kT) = kT$. Now let us examine this same system using the quantum-mechanical expression for the energy of a harmonic oscillator (ignoring the zero-point energy — see the problems)

$$\epsilon = nh\nu \qquad (n = 0, 1, 2, \ldots)$$

where h is Planck's constant and ν is the frequency of the oscillator. As discussed in Section 5 of Chapter 2, $h\nu$ typically has a value of about 10^{-13} erg. The partition function for this "system" (i.e., one harmonic oscillator) is

$$Z = \sum_{n=0}^{\infty} e^{-\beta nh\nu} = \sum_{n=0}^{\infty} (e^{-\beta h\nu})^n \tag{6.3}$$

Using the fact that

$$\frac{1}{1-x} = \sum_{n=0}^{\infty} x^n \qquad (x < 1)$$

we find (setting $x = e^{-\beta h\nu}$)

$$Z = \frac{1}{1 - e^{-\beta h\nu}} \tag{6.4}$$

Using this result, we find that the average energy of the oscillator is

$$-\frac{\partial \ln Z}{\partial \beta} = +\frac{\partial}{\partial \beta} \ln (1 - e^{-\beta h\nu}) = +\frac{h\nu}{e^{\beta h\nu} - 1}$$

Therefore, according to quantum mechanics, *the average energy of a* **harmonic oscillator** *is*

$$\bar{\epsilon} = \frac{h\nu}{e^{\beta h\nu} - 1} \tag{6.5}$$

The question obviously arises, "Under what conditions does (6.5) agree with the classical value of $\bar{\epsilon} = kT$?"

To answer this, we note first that kT equals $1.38 \times 10^{-16}T$ (erg). If the temperature is high, then this can be much larger than $h\nu$ ($\simeq 10^{-13}$ to 10^{-14} erg), so that $\beta h\nu = h\nu/kT \ll 1$. In this case Equation (6.5) reduces to

$$\bar{\epsilon} \simeq \frac{h\nu}{1 + \beta h\nu - 1} = \frac{1}{\beta} = kT \qquad (\beta h\nu \ll 1) \tag{6.6}$$

Here we have used the approximation $e^{\beta h\nu} \simeq 1 + \beta h\nu$. The result (6.6) now agrees with the prediction of the classical equipartition law. We see that the quantum and classical results agree in this case, provided that the *difference* between the energy levels of the oscillator, namely $h\nu$, is much smaller than kT. It is sometimes useful to express this fact directly in terms of temperatures. If we define the **characteristic vibrational temperature** by

$$\Theta_{\text{vib}} \equiv \frac{h\nu}{k} \tag{6.7}$$

then the classical approximation is good when $T \gg \Theta_{\text{vib}}$ (that is, $1 \gg \beta h\nu$), as can be seen from (6.6). Typically Θ_{vib} is of the order of several thousand degrees Kelvin (see problems).

In the case of low temperatures ($\beta h\nu \gg 1$ or $\Theta_{\text{vib}} \gg T$) Equation (6.5) reduces to

$$\bar{\epsilon} \simeq h\nu e^{-\beta h\nu} \qquad (\beta h\nu \gg 1) \tag{6.8}$$

This value for the average differs greatly from the equipartition value kT when T is small (see problems). This shows that the classial theory is not accurate if the energy difference between the quantum states is smaller than, or of the order of, kT. We shall see in the subsequent sections that this is a general rule, namely:

A necessary condition for classical statistical mechanics to yield accurate results is that the energy difference between the quantum states must be much less than kT. (6.9)

It should be emphasized that (6.9) gives a *necessary*, but not a sufficient, condition for classical statistical mechanics to be accurate. That is, other conditions also must be satisfied. These conditions are important, however, only if the temperature is very low or the density of particles is very high; they need not concern us for the present (see Sections 14–16). Many of the important discrepancies between the observed properties of systems and the results predicted by classical statistical mechanics are due to the fact that condition (6.9) is not satisfied. We shall consider a number of such examples in the following sections.

PROBLEMS

1. According to Equation (4.8), the average energy of a particle in a gravitational field is $(5/2)kT$. Moreover, these particles have only *three* quadratic terms in their energy. According to the equipartition law, one would expect the value of $(5/2)\,kT$ for a molecule with *five* quadratic terms in the expression for its energy. What is the resolution of this "paradox"?
2. A spring balance, which obeys Hooke's law, is used to measure a mass m. The spring is suspended vertically and has a spring constant κ (do not confuse this with Boltzmann's constant).

(a) Let z be measured downward from the equilibrium point of the end of the spring before the mass is attached. Obtain an expression for the potential energy $\phi(z)$ of the mass when attached to the spring in a gravitational field. Use $\phi(z)$ to determine the mechanical equilibrium value of z for the mass m (call it $\bar{z}$). By adding a constant to $\phi(z)$, arrange it so that $\phi(\bar{z}) = 0$.

(b) Now, if the spring, mass, and surrounding gas are in equilibrium at a temperature T, the extension of the spring will fluctuate around the value $z = \bar{z}$. Making use of the equipartition law and part (a), obtain an expression for $\overline{(z - \bar{z})^2}$.

(c) For what value of m does $\sigma(z) = \bar{z}$? What is the significance of this value of m?

3. Repeat the derivation of Equation (6.5) retaining the zero-point energy of the harmonic oscillator [so that $\epsilon = (n + \frac{1}{2})h\nu$]. Does the additional term in $\bar{\epsilon}$ depend on the temperature? Explain why this term can be ignored.

4. Show that, if the summation in Equation (6.3) is replaced by an integral over the continuous variable n, the resulting partition function yields the classical result for $\bar{\epsilon}$. The error that results from replacing a summation by an integral can be determined with the help of the Euler-Maclaurin series

$$\sum_{n=0}^{m} f(a + nb) = \int_0^m f(a + nb)\,dn + \tfrac{1}{2}[f(a + mb) + f(a)] + \frac{b}{12}[f'(a + mb) - f'(a)] + \cdots$$

where $f'(x) \equiv df/dx$, and the remaining terms contain higher powers of b, which can generally be ignored if b is small. Use this series to obtain an approximate expression for the series in Equation (6.3) (identifying m, a, and b). Demonstrate that only the integral is important if $\beta h\nu \ll 1$.

5. For a harmonic oscillator determine the ratio of $\bar{\epsilon}$ given by the quantum theory [Equation (6.5)] to that given by the classical theory, if $\beta h\nu = 0.5$, 1, or 5.

6. Using the results in Chapter 2, determine the approximate temperature at which kT is of the order of the separation of the *translational* energy levels of a He atom in a container that is 10 cm on a side. Is condition (6.8) satisfied for ordinary temperatures?

7. Using the values of ν in Table 4 of Chapter 2, determine the characteristic vibrational temperature Θ_{vib} of the diatomic molecules HCl, CO, O_2, and Cl_2.

7. *HEAT CAPACITIES OF GASES (CLASSICAL THEORY)*

One of the original applications of the classical equipartition law, and a subsequent indication of its limitations, was the prediction of the heat capacities of rarefied (perfect) gases. We shall consider these in the order of increasing complexity.

Perfect Monatomic Gases. This case has already been discussed in previous sections, but we shall include it here for completeness. In the case of a perfect monatomic gas, the classical expression for the energy of the system is

$$E = \sum_{i=1}^{N} \tfrac{1}{2} m\mathbf{v}_i^2$$

The energy, which is entirely due to translational motion, is the sum of $3N$ quadratic terms. According to the equipartition law, each quadratic term should contribute $\frac{1}{2}kT$ to the average energy of the system, so we should have $U = 3N(\frac{1}{2}kT)$. The predicted heat capacity is then

$$C_V = \left(\frac{\partial U}{\partial T}\right)_V = \frac{3Nk}{2} = \frac{3nR}{2}$$

where n is the number of moles and R is the universal gas constant. Since $C_p = C_V + nR$ for a perfect gas, this predicts that $C_p = 5nR/2$. One of the more readily measurable quantities is the ratio $\gamma = C_p/C_V$, which the classical theory predicts to be

$$\gamma = \frac{\dfrac{5nR}{2}}{\dfrac{3nR}{2}} = \frac{5}{3} = 1.667$$

Typical experimental values for γ are

$$\gamma = 1.66 \text{ (He)}; \quad \gamma = 1.64 \text{ (Ne)}; \quad \gamma = 1.67 \text{ (Ar)};$$
$$\gamma = 1.68 \text{ (Kr)}; \quad \gamma = 1.66 \text{ (Xe)}$$

These values are all in good agreement with the predicted value of $\gamma = 1.667$. We see, therefore, that the classical theory is quite adequate in this case — a result of the fact that the internal energy is due entirely to translational motion, as we shall soon see.

Perfect Diatomic Gases. In Chapter 2 we discussed the fact that the energy of a diatomic molecule is of the form

$$\epsilon = \epsilon_{\text{trans}} + \epsilon_{\text{rot}} + \epsilon_{\text{vib}} \tag{7.1}$$

corresponding to the sum of the translational, rotational, and vibrational energies. In Section 4 of Chapter 2 we considered two possible classical models of such molecules and obtained the following classical expressions for the energies:

$$\epsilon = \frac{1}{2} M\mathbf{V}^2 + \frac{1}{2} I(\omega_1^2 + \omega_2^2) \qquad \text{(rigid dumbbell model)} \tag{7.2}$$

and

$$\epsilon = \frac{1}{2}M\mathbf{V}^2 + \frac{1}{2}I(\omega_1^2 + \omega_2^2) + \frac{1}{2}\mu\left(\frac{dr_{ab}}{dt}\right)^2 + \frac{1}{2}\kappa(r_{ab} - r_0)^2 \qquad (7.3)$$

(harmonic model)

The term $\frac{1}{2}M\mathbf{V}^2$ represents the translational energy of the entire molecule (M = total mass, $\mathbf{V}$ = velocity of the center of mass). The terms $\frac{1}{2}I(\omega_1^2 + \omega_2^2)$ represent the rotational energy of the molecule (I = moment of inertia = μr_0^2, and ω_1, ω_2 are angular velocities). In the dumbbell model, it is assumed that there is no vibrational motion ($\epsilon_{\text{vib}} = 0$), whereas $\epsilon_{\text{vib}} = \frac{1}{2}\mu(dr_{ab}/dt)^2 + \frac{1}{2}\kappa(r_{ab} - r_0)^2$ is the harmonic model [μ = reduced mass = $m_a m_b/(m_a + m_b)$, κ = force constant, r_{ab} = separation distance of the two atoms, and r_0 = equilibrium separation distance]. Since the forces between the atoms are finite, there is of course no way in which a perfectly rigid dumbbell could be formed in the classical theory. Therefore, on the basis of classical mechanics, the only realistic model is the harmonic model (7.3). The dumbbell model is sort of a "fake" from the point of view of classical mechanics.

Now let us see what these classical models predict for the heat capacities of diatomic gases. The total energy of a perfect gas is given by

$$E = \sum_{i=1}^{N} \epsilon_i \qquad (7.4)$$

where ϵ_i is the energy (7.1) of the ith molecule. In the classical theory we can again apply the equipartition law to (7.4), using (7.3) for the molecular energy. We note that ϵ, as given by (7.3), consists of the sum of *seven quadratic terms* (three translational, two rotational, and two vibrational). In the vibrational term the variables are dr_{ab}/dt (the velocity along the connecting axis) and the combination $(r_{ab} - r_0)$. In this case the total energy (7.4) consists of the sum of $7N$ quadratic terms, and according to the equipartition law we should have

$$U = 7N\left(\frac{1}{2}kT\right) \qquad \text{so} \qquad C_V = \left(\frac{\partial U}{\partial T}\right)_V = \frac{7}{2}Nk$$

Since $C_p = C_V + Nk = \frac{9}{2}Nk$, the harmonic model predicts that $\gamma = C_p/C_V$ should have the value

$$\gamma = \tfrac{9}{7} = 1.286 \qquad \text{(harmonic model)} \qquad (7.5)$$

The experimental values for several diatomic gases (at 15°C and 1 atmosphere pressure) are listed below:

$$\begin{aligned} &\gamma = 1.408 \text{ (for } H_2); \quad \gamma = 1.400\ (O_2); \quad \gamma = 1.404\ (N_2); \\ &\gamma = 1.34\ (Cl_2); \qquad\quad \gamma = 1.404\ (CO) \end{aligned} \qquad (7.6)$$

We see that, with the exception of chlorine, all these values are very close to $\gamma = 1.40$. It is clear therefore that the value (7.5) predicted by the classical *harmonic* model is *not* accurate. The fact that this presumably realistic model does not yield accurate results for the heat capacity was described by Lord Kelvin, in 1900, as one of the "nineteenth-century clouds over the dynamical theory of heat." It took the development of quantum mechanics to disperse this and other "clouds" that appeared in statistical mechanics.

To see that the model of the molecule makes a significant difference in the predicted value of γ, let us consider the result obtained for the rigid dumbbell model. In this model the energy of a molecule, Equation (7.2), consists of the sum of *five* quadratic terms (rather than seven), because there is no vibrational motion. The total energy of the system (7.4) then contains $5N$ quadratic terms. According to the equipartition law, the internal energy of this system should be $5N(\frac{1}{2}kT)$, which gives

$$C_V = \left(\frac{\partial U}{\partial T}\right)_V = \tfrac{5}{2}Nk \qquad \text{and} \qquad C_p = \tfrac{7}{2}Nk$$

Therefore the dumbbell model predicts that

$$\gamma = \tfrac{7}{5} = 1.40 \qquad \text{(dumbbell model)} \tag{7.7}$$

This value agrees closely with the experimental values in Equation (7.6) (except for Cl_2!). It must be concluded, therefore, that this dumbbell model is more realistic than the harmonic model. The difficulty is, as we have already noted, that there is no way to explain, using classical mechanics, why a molecule should behave like a rigid dumbbell. In particular, it should be noted that a harmonic molecule that has a *very* large force constant (corresponding to a very stiff spring between the atoms) is *not* the same as an *absolutely* rigid dumbbell. This is because the equipartition law predicts an average energy of $\frac{1}{2}kT$ for every quadratic variable regardless of the value of the coefficient κ. Thus, on the basis of classical mechanics, there is no way to explain why the vibrational motion of diatomic molecules should be completely "frozen out." Moreover there are cases, such as Cl_2, that are somehow "not quite frozen" (i.e., γ lies between 1.400 and 1.286). Such cases are a complete mystery from the point of view of classical mechanics.

In the last section we found that the classical prediction for the average energy of a harmonic oscillator is inaccurate if $h\nu \gtrsim kT$, or in other words $\Theta_{\text{vib}} \equiv h\nu/k \gtrsim T$. Since $\Theta_{\text{vib}} \simeq 1000$ to $2000°K$, it is no great surprise that the observed values of γ for $T \simeq 300°K$ do not agree with the value obtained from the equipartition law. In effect the available energy per molecule (kT) is too small to excite the vibrational motion of most molecules — because this excitation requires an energy $h\nu \gg kT$. Since the vibrational motion is not excited, it is "frozen," and therefore the molecules behave like rigid dumbbells. The picture is quite clear; the only problem is to replace these words with some numbers. In particular, it would be nice to predict the value of γ for the "partially frozen" case of Cl_2. To do this we must consider the quantum partition function in a little more detail than we have done so far. These considerations are taken up in the next section.

PROBLEMS

1. Polyatomic molecules fall into two classes: (a) linear molecules, in which the atoms lie along a line, and (b) nonlinear molecules. The rotational energy of linear molecules is similar to that of diatomic molecules, whereas the rotational energy of nonlinear molecules is of the form $\frac{1}{2}(I_1\omega_1^2 + I_2\omega_2^2 + I_3\omega_3^2)$. In other words, if the atoms do not lie on a line (nonlinear), there are effectively three axes about which the molecule can rotate. The contribution of the vibrational motion also tends to be larger for the larger molecules (they have more ways in which to vibrate — some of which take very little energy). From the following experimental values (for $T \simeq 290°K$), determine the vibrational contribution to C_V, assuming that the rotational contribution obeys the equipartition law: (a) linear; CO_2 ($\gamma = 1.30$), C_2H_2 ($\gamma = 1.27$), and (b) nonlinear; H_2S ($\gamma = 1.34$), C_2H_6 ($\gamma = 1.22$). Note and discuss the significance of any abnormal values.
2. When a gas is not very rarefied, the interaction energy between the molecules affects the heat capacity (as well as the equation of state) of the gas. Refer to Section 5, and let $Z = Z^*Z_{\text{conf}}$, where Z_{conf} is given by (5.7) and Z^* is the remaining part in (5.4). Show that, with the approximation (5.12), the heat capacity is

$$C_V = C_V\,(\text{perfect gas}) - \left(\frac{N^2k}{V}\right)\frac{d}{dT}\left(T^2\frac{dB}{dT}\right)$$

Using the approximate expressions for $B(T)$ given in (5.17) and (5.21), determine this correction factor to C_V.

8. QUANTUM STATISTICS OF NONINTERACTING PARTICLES

Let us now consider the nature of the quantum partition function in more detail. The partition function is

$$Z = \sum_{\text{ms}} e^{-\beta E} \tag{8.1}$$

where the sum is over all possible microstates (ms) — i.e., quantum states — of the system, and E is the total energy of the system. First of all we shall restrict our discussion to the case of *noninteracting particles*, so that E is equal to the *sum of the energies* of these particles. Moreover, we shall use the term "particle" in a generic sense to represent any dynamic quantity that has a well-defined energy ϵ. Thus a "particle" may be an electron, a molecule, or even a sound wave in a solid or a light wave.

Now the first difference between (8.1) and the classical partition function is, of course, that the possible values of the energy are discrete rather than continuous. This simply means that we must evaluate sums rather than integrals. As we have already seen in Section 6, the evaluation of a sum need not be difficult.

A second difference is that in quantum mechanics we must distinguish between two possible situations. If two particles have some measurable property [such as mass, electrical charge, or (in the case of waves) frequency] that is different, then we say that the two particles are **distinguishable.** If, on the other hand, there is no experimental method of identifying one particle from another, then we say they are **indistinguishable.** Thus two identical He atoms are indistinguishable, whereas a He atom and a Ne atom are distinguishable. Similarly, two sound waves in a solid are distinguishable because of their different wavelengths or directions of propagation. This distinction between distinguishable and indistinguishable particles is one of the new concepts introduced by quantum mechanics. In classical mechanics it is always assumed that particles are distinguishable (i.e., can be separately identified) even if their physical properties are the same — the idea being that we could always "keep track of which is which" simply by measuring their position and velocity at small time intervals. Quantum mechanics, however, places restrictions on the possible accuracy of such measurements, and this effectively makes it impossible to "keep track of which is which," unless they have different physical properties.

Now what has this to do with the evaluation of the partition function Z? The summation in (8.1) is to be taken over all the different microstates of the system. By *different microstates* we now mean *distinguishably different* microscopic conditions within the system. Whether or not two situations are distinguishable now must be judged on the basis of whether or not the particles are distinguishable in the above sense. Let us consider these two possibilities separately.

Case I: Distinguishable Particles. The simplest case is that in which the "particles" of the system can be distinguished from one another by some physical property. Apparently a gas containing one type of gas molecule does *not* belong to the present case. In fact the most common situation involving N distinguishable "particles" usually refers to sound waves in solids, or electromagnetic waves — each wave ("particle") being distinguishable from another wave by its wavelength or direction of propagation. Both of these physical systems will be treated in detail in following sections.

The essential point about distinguishable particles is that we "can keep track of which is which." In this case it is legitimate to *label* the particles $i = 1, 2, \ldots, N$, and to speak of the energy ϵ_i of the ith particle. Then the total energy of N *noninteracting* particles can be written

$$E = \sum_{i=1}^{N} \epsilon_i \tag{8.2}$$

and the partition function is therefore

$$Z = \sum_{\text{ms}} e^{-\beta \Sigma \epsilon_i} \tag{8.3}$$

The summation over the microstates now is simply the sum over all possible values of $\epsilon_1, \epsilon_2, \ldots, \epsilon_N$. Therefore (8.3) becomes

$$Z = \left(\sum_{\epsilon_1} e^{-\beta\epsilon_1}\right)\left(\sum_{\epsilon_2} e^{-\beta\epsilon_2}\right) \times \cdots \times \left(\sum_{\epsilon_N} e^{-\beta\epsilon_N}\right) \tag{8.4}$$

which also can be written

$$Z = \prod_{i=1}^{N} \left(\sum_{\epsilon_i} e^{-\beta\epsilon_i}\right) \tag{8.4}$$

where Π represents the *product* (in the same way as Σ represents the sum). Until we are given some information about the possible values of $\epsilon_1, \epsilon_2, \ldots, \epsilon_N$, we cannot do anything further with the expression (8.4) — so let us consider the second possibility.

Case II: Indistinguishable Particles. In quantum mechanics the point of view is taken that, if particles are physically indistinguishable from one another, then it makes no sense to label the particles $i = 1, 2, \ldots, N$ (as we did in case I). If you *could* assign numbers to each particle, this would imply that you *could* distinguish between them, and quantum mechanics rejects this possibility for physically identical particles.

Since we cannot label the particles, how do we characterize a microstate of the system? Let $\epsilon(s)$ represent the possible values of the energy of any *one* of the identical particles [i.e., each particle can have the energy values $\epsilon(1), \epsilon(2), \epsilon(3), \ldots$]. Note that s does *not* refer to the particle, but rather to the particle's *quantum state* [in case I, ϵ_i represented the energy of the ith particle, whereas $\epsilon(s)$ represents the energy of the sth *quantum state* of *any* particle]. Now, since we cannot label the particles, the only thing we can do is to *count* the *number* of particles in each of the quantum states s. This leads to the concept of **occupation numbers** n_s. Let

$$n_s = \textit{the number of particles in the particle state } s \tag{8.5}$$

We can distinguish between two microscopic situations in the system only if these occupation numbers have different values in the two cases. Therefore we conclude that

For a system of noninteracting indistinguishable particles, a microstate is specified by the values of all the occupation numbers $(n_1, n_2, n_3, \ldots)$. (8.6)

For example, if a system had particles with energies $\epsilon(3)$, $\epsilon(1)$, $\epsilon(3)$, $\epsilon(4)$, $\epsilon(2)$, and $\epsilon(1)$, then it would be in the microstate $(2, 1, 2, 1, 0, \ldots)$. In terms of these occupation numbers, the total energy of the system of noninteracting particles is simply

$$E = \sum_{s=1}^{\infty} n_s \epsilon(s) \tag{8.7}$$

(i.e., the number of particles in the particle state s, times the energy of that state, summed over all states). Moreover, if there are N particles, it should be clear that, since each particle is in some particle state, we must have

$$\sum_{s=1}^{\infty} n_s = N \tag{8.8}$$

With this out of the way, let us take another look at the partition function

$$Z = \sum_{\text{ms}} e^{-\beta E} \tag{8.1}$$

Presumably what we should now do is to substitute the value of E, given by (8.7), into this equation and then sum over all allowed values of $(n_1, n_2, n_3, \ldots)$ that satisfy (8.8). For example, the microstate $(1, 0, 2, 1, 0, \ldots)$ would contribute the additive term $e^{-\beta[\epsilon(1)+2\epsilon(3)+\epsilon(4)]}$ to Z (what is N in this example?). According to (8.6) such a summation represents the summation over all microstates. Unfortunately this summation is *not* easy to do, and we shall defer a discussion of this general case to a later section (Section 14).

In many situations it is sufficient to use a very useful approximation, which can be treated fairly easily. The simplification occurs when a system is **nondegenerate.** By a *nondegenerate system* we mean one that satisfies the following condition.

$$[\text{The number of particle states with } \epsilon(s) < kT] \gg N \tag{8.9}$$

This is illustrated schematically in Figure 8.1. The probability that a particle is in a particle state with energy $\epsilon(s)$ is proportional to $e^{-\beta\epsilon(s)}$. If there are many states with $\epsilon(s) < kT$, then a particle is equally likely to be in any of these states. When (8.9) is satisfied, each particle has a large number of states to choose from, each with about the same probability. Moreover, since the number of these states is much greater than N, it is

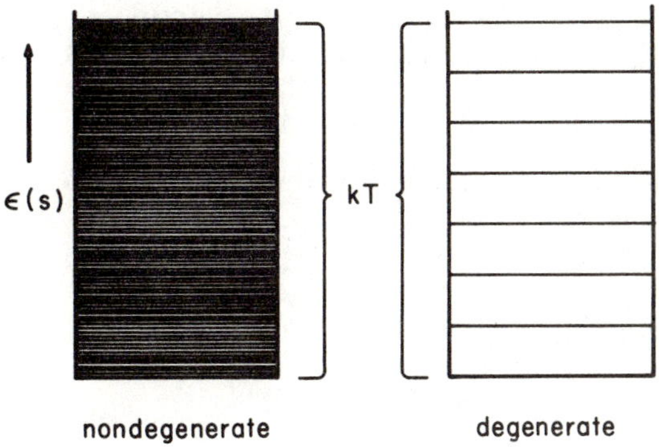

Figure 8.1

unlikely for two particles to be in the same state. In terms of the occupation numbers n_s this means that for most *microstates either* $n_s = 0$ *or* $n_s = 1$ (*for all s*). First of all let us see how this fact simplifies the evaluation of Z, and then we shall return to the question of when (8.9) is satisfied [it should be clear, however, that (8.9) can always be satisfied if T is sufficiently large, and N is not too large].

Consider first a system with only one particle ($N = 1$). In this case the microstates are (1, 0, . . .), (0, 1, 0, . . .), and so on (corresponding to $n_1 = 1$, *or* $n_2 = 1$, *or* $n_3 = 1$, and so on). The partition function is therefore

$$Z(N = 1) \equiv \mathfrak{z} = e^{-\beta\epsilon(1)} + e^{-\beta\epsilon(2)} + e^{-\beta\epsilon(3)} + \cdots$$

or

$$\mathfrak{z} = \sum_{s=1}^{\infty} e^{-\beta\epsilon(s)} \tag{8.10}$$

which is often referred to as the **particle partition function.**

Next, consider the case $N = 2$. In this case some microstates correspond to both particles' being in the same particle state [for example, (0, 2, 0, . . .)]. However, if the system is *nondegenerate*, then most of the microstates will correspond to the two particles' being in different particle states (so $n_s = 0$ or $n_s = 1$). If we ignore the microstates for which $n_s = 2$, then Z is *approximately* given by

$$\begin{aligned} Z(N = 2) \simeq\ & e^{-\beta[\epsilon(1)+\epsilon(2)]} + e^{-\beta[\epsilon(1)+\epsilon(3)]} + e^{-\beta[\epsilon(1)+\epsilon(4)]} + \cdots \\ & + e^{-\beta[\epsilon(2)+\epsilon(3)]} + e^{-\beta[\epsilon(2)+\epsilon(4)]} + e^{-\beta[\epsilon(2)+\epsilon(s)]} + \cdots \end{aligned} \tag{8.11}$$

Now we note that if we take the expression (8.10) for $\mathfrak{z}$ and square it, then we obtain an expression very similar to (8.11). That is

$$\begin{aligned} \mathfrak{z}^2 = (e^{-\beta\epsilon(1)} + e^{-\beta\epsilon(2)} + \cdots)^2 = \underline{e^{-\beta 2\epsilon(1)}} + 2e^{-\beta[\epsilon(1)+\epsilon(2)]} \\ + 2e^{-\beta[\epsilon(1)+\epsilon(3)]} + \cdots + \underline{e^{-\beta 2\epsilon(2)}} + 2e^{-\beta[\epsilon(2)+\epsilon(3)]} \\ + 2e^{-\beta[\epsilon(2)+\epsilon(4)]} + \cdots \end{aligned}$$

If we again ignore the terms corresponding to $n_s = 2$ (the underlined terms), we see that the last expression and (8.11) differ only by a factor of two, so

$$Z(N = 2) \simeq \tfrac{1}{2}\mathfrak{z}^2$$

If we consider larger values of N, we obtain a similar result. For $N = 3$, Z contains terms such as

$$e^{-\beta[\epsilon(1)+\epsilon(2)+\epsilon(3)]}$$

whereas $\mathfrak{z}^3$ yields terms of the form

$$3!\, e^{-\beta[\epsilon(1)+\epsilon(2)+\epsilon(3)]}$$

To make $Z(N = 3)$ approximately equal to $\mathfrak{z}^3$, we must divide $\mathfrak{z}^3$ by 3!, so

$$Z(N = 3) \simeq \frac{1}{3!}\mathfrak{z}^3$$

This can be extended immediately to larger values of N, and we find that for *indistinguishable particles*

$$\boxed{Z \simeq \frac{1}{N!}\mathfrak{z}^N \qquad \text{(nondegenerate system)}} \tag{8.12}$$

This *approximate* result for *indistinguishable particles* is also related in a simple way to the partition function we obtained for *distinguishable* particles:

$$Z = \prod_{i=1}^{N}\left(\sum_{\epsilon_i} e^{-\beta\epsilon_i}\right) \tag{8.4}$$

If the possible values of ϵ_i for *all* the particles were the same, say $\epsilon(1)$, $\epsilon(2)$, $\epsilon(3)$, . . . , then each of the sums in (8.4) would be identical and equal to $\sum_{s=1}^{\infty} e^{-\beta\epsilon(s)}$. Since (8.4) contains a product of N of these sums, we would have

$$Z = \left[\sum_{s=1}^{N} e^{-\beta\epsilon(s)}\right]^N = \mathfrak{z}^N \tag{8.13}$$

where the last equality comes from the definition of $\mathfrak{z}$, (8.10). We see that the result (8.13) for *distinguishable* particles differs from the *approximate* result (8.12) for *indistinguishable* particles by a factor of $(1/N!)$. This difference is due to the difference in the number of microstates in the two cases. We can easily understand this by taking a simple example. In the accompanying table we have listed six microstates for a system with three *distinguishable* particles. Note that these microstates are *nondegenerate* (no two particles have the same energy). Since the particles can be labeled in

	Particle		
	1	2	3
ms 1	$\epsilon(2)$	$\epsilon(5)$	$\epsilon(3)$
ms 2	$\epsilon(2)$	$\epsilon(3)$	$\epsilon(5)$
ms 3	$\epsilon(3)$	$\epsilon(2)$	$\epsilon(5)$
ms 4	$\epsilon(5)$	$\epsilon(2)$	$\epsilon(3)$
ms 5	$\epsilon(5)$	$\epsilon(3)$	$\epsilon(2)$
ms 6	$\epsilon(3)$	$\epsilon(5)$	$\epsilon(2)$

this case, we have $3 \times 2 = 3!$ microstates. For a system of *indistinguishable* particles, however, there would be only *one* microstate with these energies — with occupation numbers (0, 1, 1, 0, 1, 0, . . .). Therefore, when you label the particles (distinguishable) you have $N!$ more *nondegenerate* microstates than in the case of indistinguishable particles. The factor $N!$ comes from the number of ways one can distribute N different energies $\epsilon(s)$ over N distinguishable particles. This is the reason for the factor of $N!$ difference between (8.13) and (8.12).

Before we shall apply these results, let us determine under what conditions the inequality

$$[\text{The number of particle states with } \epsilon(s) < kT] \gg N \tag{8.9}$$

is satisfied for a *perfect gas*. That is, we will determine when a noninteracting group of gas molecules is nondegenerate. The quantum expression for the translational energy ϵ_{trans} for a molecule in a cubical container with sides of length L is given by [Chapter 2, Equation (40)]

$$\epsilon_{\text{trans}} = \frac{h^2}{8mL^2}(n_x^2 + n_y^2 + n_z^2) \equiv \frac{h^2}{8mL^2}\mathbf{n}^2 \tag{8.14}$$

where the translational quantum numbers (n_x, n_y, n_z) are all positive integers. Keep in mind that the difference between the energy levels is generally very small (Chapter 2, Section 5). According to (8.9), the system will not be degenerate if the number of states with $\epsilon_{\text{trans}} < kT$ is much larger than N. How many translational states are there that satisfy this condition? Let us set

$$(\epsilon_{\text{trans}})_{\max} = \frac{h^2}{2mL^2}\mathbf{n}_{\max}^2 = kT \tag{8.15}$$

and consider a coordinate system (n_x, n_y, n_z) shown in Figure 8.2. The possible values of (n_x, n_y, n_z) are all *positive integers*. Hence the allowed values of (n_x, n_y, n_z) are represented by points in this space which are arranged in a cubic lattice. Each cube has sides of "length" one, and hence

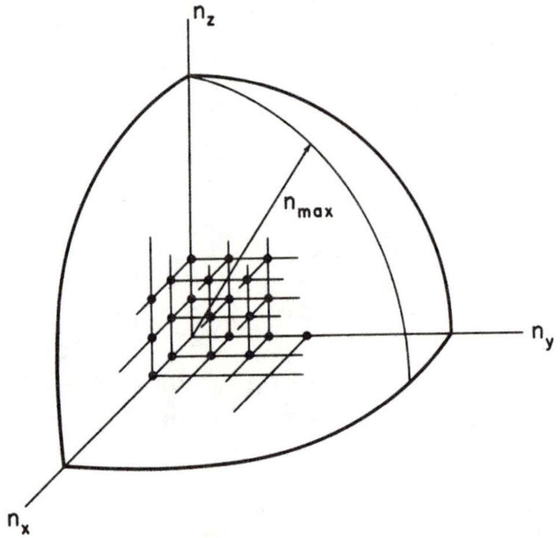

Figure 8.2

a volume of one. Moreover there is one such cube for each allowed point. The number of states that satisfy $\epsilon_{\text{trans}} \leq kT$ equals the number of points that are inside the spherical region of radius n_{max}, shown in Figure 8.2. Since this spherical region is just one-eighth of a complete sphere [because all (n_x, n_y, n_z) are positive], it has a "volume"

$$(\tfrac{1}{8})(\tfrac{4}{3})\pi n_{\text{max}}^3$$

and since the "volume" per point is unity, the number of states inside this sphere is approximately $(\frac{1}{8})(\frac{4}{3})\pi n_{\text{max}}^3/1$. Therefore, in this case, condition (8.9) becomes

$$(\tfrac{1}{8})(\tfrac{4}{3})\pi n_{\text{max}}^3 \gg N \tag{8.16}$$

From (8.15) we have $n_{\text{max}} = L(2mkT/h^2)^{1/2}$, and if we substitute this into (8.16), we obtain the following condition for a perfect gas to be *nondegenerate:*

$$\frac{\pi}{6} L^3 \left(\frac{2mkT}{h^2}\right)^{3/2} \gg N$$

or, since $L^3 = V$, the volume of the container, we have

$$\boxed{\frac{\pi}{6}\left(\frac{2mkT}{h^2}\right)^{3/2} \gg \frac{N}{V} \qquad \text{(nondegenerate perfect gas)}} \tag{8.17}$$

Apparently a perfect gas is more likely to be degenerate if (a) the molecular mass m is small, (b) the temperature is low, and/or (c) the number density N/V is large. For example, for He ($m = 4 \times 1.66 \times 10^{-24} = 6.64 \times$

10^{-24} gm) at $T = 300°K$, $2mkT/h^2 = 2 \times 6.64 \times 10^{-24} \times 1.38 \times 10^{-16} \times 300/(6.62 \times 10^{-27})^2 = 1.25 \times 10^{16}$. Therefore the system is nondegenerate if $N/V \ll 7.3 \times 10^{23}$ cm^{-3}. A density of this magnitude would correspond to a pressure of roughly 30,000 atmospheres (if $T = 300°K$)! Therefore, *at ordinary pressures, perfect gases are not degenerate.* The reason is that the energy difference between the translational states is usually very small (compared to kT). In fact there are many translational states that have the *same energy* (see problems). Therefore, if the number of particles is not too large, they are all likely to be in different particle states (i.e., nondegenerate). A counterexample is given in the problems.

PROBLEMS

1. A system contains four noninteracting particles with the energies $\epsilon(1)$, $\epsilon(3)$, $\epsilon(7)$, and $\epsilon(3)$. How many microstates are there that satisfy this requirement if (a) the particles are indistinguishable, (b) the particles are distinguishable? In each case list all of the possible microstates by a suitable method (see text). (c) Are these microstates degenerate or nondegenerate? Does the ratio of the number of microstates in (b) to the number in (a) equal $N!$? Does this conflict with the results found in this section? Explain.

2. A system contains two particles, each of which can only have the energy $\epsilon(1)$, $\epsilon(2)$, or $\epsilon(3)$. Write out all of the terms of the partition function if the particles are (a) distinguishable and (b) indistinguishable. Does $Z_{\text{dist}} = \frac{1}{2}Z_{\text{indist}}$? Does either of these partition functions equal $\mathfrak{z}^2$?

3. A hypothetical system contains 10^{19} particles. The energies of the particle states are given by

$$\epsilon(s) = s\epsilon_0 \qquad (s = 1, 2, 3, \ldots)$$

where $\epsilon_0 = 4 \times 10^{-33}$ erg. Determine whether this system is degenerate or nondegenerate if $T = 300°K$.

4. Assume that the free electrons in a solid can be treated as a perfect gas (i.e., as if there were no forces between the electrons). Using a typical value of 10^{22} cm^{-3} for the number density of free electrons in a metal, determine whether or not this "electron gas" is degenerate at $T = 300°K$.

5. Indicate a group of points in Figure 8.2 (particle states) that have the same (or nearly the same) energy. From a classical point of view, what is the difference between these different translational states (in terms of the velocity of the particle)? Refer back to the discussion in Chapter 2, Section 5.

6. The classical partition function for a system of noninteracting identical particles is $Z = \mathfrak{z}^N$ [Chapter 3, Equation (23), and also problem 5].

   (a) Does the classical treatment assume that the particles are distinguishable or indistinguishable? Explain.

   (b) In order to "correct" this classical treatment, what dependency on N would you have to assign to D_N in Equation (13b) of Chapter 3?

(c) For large values of N, $N!$ can be approximated with the aid of *Stirling's approximation*

$$N! \approx (2N)^{1/2}\left(\frac{N}{e}\right)^N \qquad \text{or} \qquad \ln(N!) \approx N \ln N - N$$

(the symbol $\approx$ means that the ratio of the expressions on either side approaches unity as N goes to infinity). With the aid of this approximation, show that the expression for D_N in part (b) satisfies part (b) of problem 8, Chapter 3.

9. NONDEGENERATE PERFECT GASES

At the end of the last section it was shown that perfect gases are usually not degenerate (except when the temperature is *very* low, or the density is *very* large). In this case the approximation (8.12) can be used for the partition function

$$Z \simeq \frac{1}{N}\, \mathfrak{z}^N \tag{9.1}$$

where $\mathfrak{z}$ is the particle partition function [Equation (8.10)]

$$\mathfrak{z} = \sum_s e^{-\beta\epsilon(s)} \tag{9.2}$$

and the sum is over all the quantum states of a gas molecule. The energy ϵ of a molecule consists of the sum of the translational, rotational, and vibrational energies (and possibly electronic energies — see problems).

$$\epsilon = \epsilon_{\text{trans}} + \epsilon_{\text{rot}} + \epsilon_{\text{vib}} \tag{9.3}$$

Therefore, the quantum state of a molecule requires the specification of the values of the translational, rotational, and vibrational quantum numbers. Substituting (9.3) into (9.2), we can write $\mathfrak{z}$ in the form

$$\begin{aligned} \mathfrak{z} &= \sum_{\text{trans}} \sum_{\text{rot}} \sum_{\text{vib}} e^{-\beta(\epsilon_{\text{trans}} + \epsilon_{\text{rot}} + \epsilon_{\text{vib}})} \\ &= \sum_{\text{trans}} e^{-\beta\epsilon_{\text{trans}}} \sum_{\text{rot}} e^{-\beta\epsilon_{\text{rot}}} \sum_{\text{vib}} e^{-\beta\epsilon_{\text{vib}}} \\ &\equiv \mathfrak{z}_{\text{trans}}\, \mathfrak{z}_{\text{rot}}\, \mathfrak{z}_{\text{vib}} \end{aligned} \tag{9.4}$$

We see that the particle partition function can be written as the product of three terms, which depend respectively on the translational, rotational, and vibrational quantum states of a molecule. If we take the logarithm of the partition function (9.1) we find that

$$\ln Z \simeq N \ln \mathfrak{z} - \ln N!$$

and, if we use (9.4) for $\mathfrak{z}$, this becomes

$$\ln Z \simeq N(\ln \mathfrak{z}_{\text{trans}} + \ln \mathfrak{z}_{\text{rot}} + \ln \mathfrak{z}_{\text{vib}}) - \ln N! \tag{9.5}$$

The result of all this is that the translational, rotational, and vibrational motion each contribute an *additive term* to ln Z. Since all the thermodynamic state functions are expressed in terms of ln Z, the three forms of motion also each make an *additive* contribution to the state functions. For example, the internal energy is given by

$$\begin{aligned} U &= -\frac{\partial}{\partial\beta}\ln Z = -N\frac{\partial}{\partial\beta}\ln \mathfrak{z}_{\text{trans}} - N\frac{\partial}{\partial\beta}\ln \mathfrak{z}_{\text{rot}} - N\frac{\partial}{\partial\beta}\ln \mathfrak{z}_{\text{vib}} \\ &\equiv U_{\text{tran}} + U_{\text{rot}} + U_{\text{vib}} \end{aligned} \tag{9.6}$$

which is a sum of three terms — one due to each of the three forms of motion of a molecule. A similar result also holds for the other thermodynamic state functions. Now let us examine these three partition functions and determine their contribution to the internal energy (9.6). These results can then be compared with the results of the classical theory obtained in Section 7.

For the translational part of the partition function we use the expression for the translational energy

$$\epsilon_{\text{trans}} = \frac{h^2}{8mL^2}(n_x^2 + n_y^2 + n_z^2)$$

for a molecule of mass m in a cubical container with sides of length L. The translational quantum numbers (n_x, n_y, n_z) are all positive integers. Therefore,

$$\begin{aligned} \mathfrak{z}_{\text{trans}} &= \sum_{n_x=0}^{\infty}\sum_{n_y=0}^{\infty}\sum_{n_z=0}^{\infty} e^{-\beta h^2(n_x^2+n_y^2+n_z^2)/8mL^2} \\ &= \sum_{n_x=0}^{\infty} e^{-\beta h^2 n_x^2/8mL^2} \sum_{n_y=0}^{\infty} e^{-\beta h^2 n_y^2/8mL^2} \sum_{n_z=0}^{\infty} e^{-\beta h^2 n_z^2/8mL^2} \end{aligned}$$

Each of these summations has the same value (they are each summed over identical values), so we can write this as

$$\mathfrak{z}_{\text{trans}} = \left(\sum_{n=0}^{\infty} e^{-\beta h^2 n^2/8mL^2}\right)^3$$

Now we can again make use of the fact that the translational energy levels are very close together (that is, $\beta h^2/8mL^2 \ll 1$) and replace this summation by an integral

$$\mathfrak{z}_{\text{trans}} \simeq \left(\int_0^{\infty} dn\, e^{-\beta h^2 n^2/8mL^2}\right)^3 = \left[\frac{1}{2}\left(\frac{8mL^2\pi}{\beta h^2}\right)^{1/2}\right]^3$$

Here we have used the value for the Gaussian integral G_0 to obtain the last expression. Finally, if we use the fact that the volume of the gas is $V = L^3$, this can be written as

$$\boxed{\mathfrak{z}_{\text{trans}} = V\left(\frac{2\pi m}{\beta h^2}\right)^{3/2}} \tag{9.7}$$

If the gas molecules are *monatomic*, so that they only have translational energy, then from (9.6)

$$U = U_{\text{trans}} = -N\frac{\partial}{\partial\beta}\ln \mathfrak{z}_{\text{trans}} = \frac{3N}{2\beta} = \tfrac{3}{2}NkT$$

which agrees with our previous classical result (Section 7).

Next let us look at $\mathfrak{z}_{\text{rot}}$. For a *diatomic molecule* the possible values of ϵ_{rot} are given by (Chapter 2, Section 5)

$$\epsilon_{\text{rot}} = \left(\frac{h^2}{8\pi^2 I}\right)J(J+1) \qquad (J = 0, 1, 2, \ldots)$$

where I is the moment of inertia of the molecule about its center of mass and J is the rotational quantum number. Moreover, for each value of J there are $(2J + 1)$ quantum states, which correspond to the possible orientations of the angular momentum with respect to a fixed axis. Hence

$$\mathfrak{z}_{\text{rot}} = \sum e^{-\beta\epsilon_{\text{rot}}} = \sum_{J=0}^{\infty}(2J+1)e^{-\beta h^2 J(J+1)/8\pi^2 I} \tag{9.8}$$

where the factor $(2J + 1)$ comes from the summation over the various orientations (all having the same energy). Now there is a strong temptation to replace this summation by an integral — as we did in the case of $\mathfrak{z}_{\text{trans}}$. To examine this possibility it is useful to introduce a **characteristic rotational temperature**

$$\Theta_{\text{rot}} = \frac{h^2}{k8\pi^2 I} \tag{9.9}$$

so that (9.8) can be written

$$\mathfrak{z}_{\text{rot}} = \sum_{J=0}^{\infty}(2J+1)e^{-J(J+1)\Theta_{\text{rot}}/T} \tag{9.10}$$

Now this summation may be replaced by an integral, provided that the successive terms do not differ appreciably in magnitude (see problem 4 in Section 6). This will be the case provided that $\Theta_{\text{rot}}/T \ll 1$. Therefore, if $\Theta_{\text{rot}}/T \ll 1$, then

$$\mathfrak{z}_{\text{rot}} \simeq \int_0^{\infty} dJ\,(2J+1)e^{-J(J+1)\Theta_{\text{rot}}/T}$$

which can easily be integrated by introducing the variable

$$x = J(J + 1) \qquad \text{so} \qquad dx = (2J + 1)\, dJ$$

We find that

$$\mathfrak{z}_{\text{rot}} \simeq \int_0^\infty dx\, e^{-x\Theta_{\text{rot}}/T} = \frac{T}{\Theta_{\text{rot}}}$$

So, for *diatomic molecules*

$$\boxed{\mathfrak{z}_{\text{rot}} \simeq \frac{8\pi^2 I}{\beta h^2} \qquad \text{if } \Theta_{\text{rot}} \ll T} \tag{9.11}$$

Substituting this result into (9.6) shows that the contribution of the rotational motion to the internal energy is

$$U_{\text{rot}} = -N \frac{\partial}{\partial \beta} \ln \mathfrak{z}_{\text{rot}} = \frac{N}{\beta} = NkT \tag{9.12}$$

which again agrees with the classical equipartition result of Section 7. However, if the temperature is less than, or of the order of, the characteristic rotational temperature (so $\Theta_{\text{rot}}/T \gtrsim 1$), then (9.11) is not an accurate approximation and the classical result (9.12) no longer holds. The characteristic rotational temperatures of various molecules can be determined from (9.9). Several values are listed in Table 9.1. With the exception of

Table 9.1

Molecule	H_2	HCl	N_2	CO	O_2	Cl_2	I_2
Θ_{rot} (°K)	85.4	15.2	2.87	2.78	2.07	0.348	0.054

hydrogen, we see that all of the values of Θ_{rot} correspond to very low temperatures. Therefore, at ordinary temperatures, we expect that the classical result (9.12) should be quite accurate. The experimental values discussed in Section 7 did indeed agree with this classical result for diatomic molecules. The situation for polyatomic molecules is considerably more involved and generally does not agree with the classical theory, but we shall not consider this rather advanced topic.

Finally, let us turn to the evaluation of the vibrational term $\mathfrak{z}_{\text{vib}}$ and see whether we can explain the discrepancy between the observed and classical prediction for the heat capacity of some diatomic molecules. We proceed by using the quantum expression for the vibrational energy of a **harmonic oscillator** (Chapter 2, Section 5):

$$\epsilon_{\text{vib}} = h\nu(n + \tfrac{1}{2}) \qquad (n = 0, 1, 2, \ldots) \tag{9.13}$$

where n is the vibrational quantum number and ν is the characteristic frequency of the molecule. Using (9.13), we obtain for $\mathfrak{z}_{\text{vib}}$

$$\mathfrak{z}_{\text{vib}} = \sum_{n=0}^{\infty} e^{-\beta h\nu[n+(1/2)]} = e^{-\beta h\nu/2} \sum_{n=0}^{\infty} (e^{-\beta h\nu})^n$$

Using the fact that

$$\sum_{n=0}^{\infty} x^n = \frac{1}{1-x} \qquad (\text{if } x < 1)$$

we see that the summation in $\mathfrak{z}_{\text{vib}}$ can be written (by setting $x = e^{-\beta h\nu}$)

$$\boxed{\mathfrak{z}_{\text{vib}} = \frac{e^{-\beta h\nu/2}}{1 - e^{-\beta h\nu}}} \tag{9.14}$$

This can also be written in terms of the hyperbolic function $\sinh(x) = \frac{1}{2}(e^x - e^{-x})$:

$$\boxed{\mathfrak{z}_{\text{vib}} = \frac{2}{\sinh(\beta h\nu/2)}} \tag{9.14}$$

Just as in the rotational case, it is also useful here to introduce the concept of a **characteristic vibrational temperature**

$$\Theta_{\text{vib}} = \frac{h\nu}{k} \tag{9.15}$$

(which was discussed briefly in Section 6). Using (9.15) in (9.14), we can also write $\mathfrak{z}_{\text{vib}}$ as

$$\boxed{\mathfrak{z}_{\text{vib}} = \frac{2}{\sinh(\Theta_{\text{vib}}/2T)}} \tag{9.16}$$

Some typical examples of the value of Θ_{vib} for various diatomic molecules are given in Table 9.2. As can be seen from this table, the characteristic

Table 9.2

Molecule	H_2	HCl	N_2	CO	O_2	Cl_2	I_2
Θ_{vib} (°K)	6210	4150	3340	3100	2240	806	310

vibrational temperatures are generally quite high. Therefore, at ordinary temperature one often has $\Theta_{\text{vib}}/T \gg 1$, in which case the quantum-mechanical effects become important. To show this, consider the contribution of the vibrational motion to the internal energy:

$$U_{\text{vib}} = -N \frac{\partial}{\partial \beta} \ln \mathfrak{z}_{\text{vib}} = -N \frac{\partial}{\partial \beta} \ln \left[\frac{2}{\sinh (\beta h\nu/2)} \right]$$

where we have used (9.16) and (9.15). This equals

$$U_{\text{vib}} = +N \frac{\partial}{\partial \beta} \ln \left[\frac{\sinh (\beta h\nu/2)}{2} \right] = \frac{1}{2} h\nu N \frac{\cosh (\beta h\nu/2)}{\sinh (\beta h\nu/2)}$$

or

$$\boxed{U_{\text{vib}} = \frac{1}{2} h\nu N \coth \left(\frac{\beta h\nu}{2} \right) = \frac{1}{2} h\nu N \coth \left(\frac{\Theta_{\text{vib}}}{2T} \right)} \tag{9.17}$$

The heat capacity of the system is given by

$$\begin{aligned} C_V = \left(\frac{\partial U}{\partial T} \right)_V &= \left(\frac{\partial U_{\text{trans}}}{\partial T} \right)_V + \left(\frac{\partial U_{\text{rot}}}{\partial T} \right)_V + \left(\frac{\partial U_{\text{vib}}}{\partial T} \right)_V \\ &= (C_V)_{\text{trans}} + (C_V)_{\text{rot}} + (C_V)_{\text{vib}} \end{aligned} \tag{9.18}$$

We have already shown that the classical theory predictions of $(C_V)_{\text{trans}} = 3Nk/2$ and $(C_V)_{\text{rot}} = Nk$ agree with the present quantum theory at all normal temperatures. Now we consider the contribution $(C_V)_{\text{vib}}$, which the classical theory predicts to be Nk. According to (9.17) and (9.18)

$$(C_V)_{\text{vib}} = \frac{\partial U_{\text{vib}}}{\partial T} = \frac{\partial U_{\text{vib}}}{\partial \beta} \frac{\partial \beta}{\partial T} = -\frac{1}{kT^2} \frac{\partial U_{\text{vib}}}{\partial \beta} = \frac{1}{kT^2} \left(\frac{1}{2} h\nu \right)^2 N \operatorname{csch}^2 \left(\frac{h\nu\beta}{2} \right)$$

or

$$\boxed{(C_V)_{\text{vib}} = \frac{Nk\left(\frac{\beta h\nu}{2} \right)^2}{\sinh^2 \left(\frac{\beta h\nu}{2} \right)} = \frac{Nk\left(\frac{\Theta_{\text{vib}}}{2T} \right)^2}{\sinh^2 \left(\frac{\Theta_{\text{vib}}}{2T} \right)}} \tag{9.19}$$

The functions U_{vib} and $(C_V)_{\text{vib}}$ are shown in Figures 9.1 and 9.2, respectively. The dashed curve in these figures represents the prediction of the classical equipartition law. Since usually $\Theta_{\text{vib}} > 2000°\text{K}$ (Table 9.2), room temperature corresponds to $T/\Theta_{\text{vib}} < 0.15$. This point is indicated in the figures. We see that in this region the vibrational motion makes *no contribution* to the heat capacity of the gas. Physically what is happening is that, if $\Theta_{\text{vib}}/T = h\nu/kT \gg 1$, then the energy difference between the vibrational states ($h\nu$) is much greater than the available thermal energy per molecule (kT), so that the vibrational motion cannot be excited. In other words, the vibrational motion is "frozen out" at temperatures below Θ_{vib}. This ex-

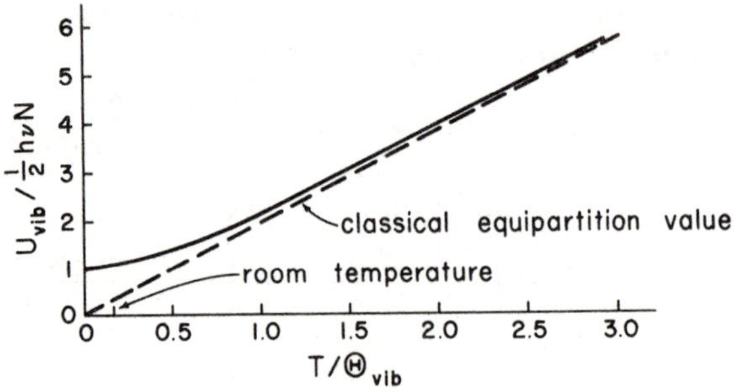

Figure 9.1

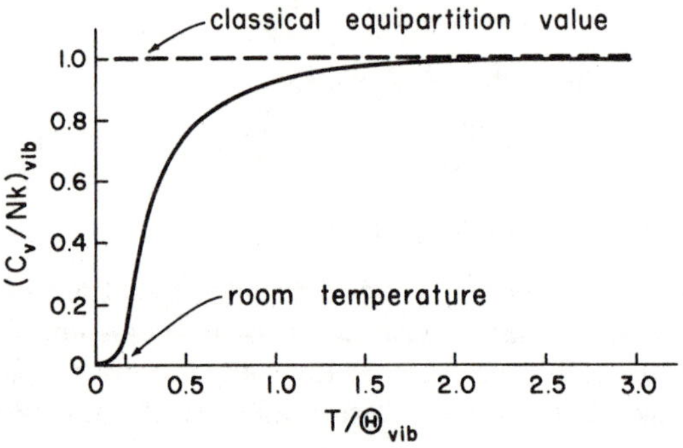

Figure 9.2

plains the experimental fact that the heat capacities of diatomic gases at room temperature are usually found to be near $C_V = (C_V)_{\text{trans}} + (C_V)_{\text{rot}} = \frac{3}{2}Nk + Nk = \frac{5}{2}Nk$ (see the values in Section 7). One of the interesting exceptions to this rule is the case of chlorine gas. As can be seen in Table 9.2, Cl_2 has a relatively low value of Θ_{vib}, and consequently the vibrational motion is only "partially frozen out" at room temperatures. Therefore, in this case, the vibrational motion *does* make a significant contribution to the heat capacity, even at ordinary temperature. Numerical examples of the agreement between the present theory and experimental values are given in the problems.

In closing, we see that quantum mechanics, with its finite separation of allowed energies, resolved one of the perplexing problems of classical statistical mechanics. An important lesson here is that even an observation that may appear to be rather "unexciting" — such as Kelvin's "black cloud" — may have its resolution in a very profound change in our comprehension of physical laws. A ripple may be the precursor of a wave.

PROBLEMS

1. In Equation (9.3) the possible contribution due to the bound electrons has been neglected. Such a constant term in the energy can be neglected as long as the electrons do not change their energy. Let $\Delta\epsilon$ represent the energy difference between two electronic states. Define an appropriate characteristic electron temperature Θ_{elec}. Determine its value if $\Delta\epsilon = 3 \times 10^{-12}$ erg ($\simeq 2$ electron volts). Will the bound electrons contribute to C_V if T is less than 2000°K?

2. As noted in Section 7, the experimental value of $\gamma = C_p/C_V = 1.34$ for Cl_2 (at 15°C) is impossible to explain on the basis of classical statistical mechanics (one example of Kelvin's "black cloud"). Use the data in Tables 9.1 and 9.2 together with the theory of this section to predict the value of γ for Cl_2 at 15°C. Note that $C_p = C_V + Nk$ is a *thermodynamic* relationship for perfect gases (i.e., it is independent of the statistical theory).

3. Nitric oxide (NO) is an exception to the results of problem 1, because the energy difference $\Delta\epsilon$ between the lowest and first excited electronic state is only $\Delta\epsilon = 2.4 \times 10^{-14}$ erg. Determine $\Theta_{\text{elec}} \equiv \Delta\epsilon/k$ for this gas. Using the fact that there are two lowest electronic states ($\epsilon_{\text{elec}} = 0$) and two electronic states with $\epsilon_{\text{elec}} = \Delta\epsilon$, write out the four terms of $\mathfrak{z}_{\text{elec}}$ (neglecting the higher energy electronic states, which contribute very little). Obtain an expression for $(C_V)_{\text{elec}}$ in terms of Θ_{elec}, and evaluate it for $T = 298$°K. Using $\Theta_{\text{vib}} = 2740$°K and $\Theta_{\text{rot}} = 2.5$°K, predict the value of C_V for NO at $T = 298$°K.

4. The experimental value of C_V for N_2 at different temperatures has been found to be

T (°K)	170	500	770	1170	1600	2000	2440
C_V/Nk	2.50	2.57	2.76	3.01	3.22	3.31	3.40

Compare these values with the theoretical predictions of this section.

10. HEAT CAPACITIES OF SOLIDS

As we saw in discussing the heat capacities of perfect gases, the classical theory of statistical mechanics cannot predict many of the observed features of the heat capacities, particularly at low temperatures. A similar difficulty arises in predicting the heat capacities of solids. Not only do the observed heat capacities differ from the classical equipartition results, even at fairly high temperatures (say 200°K), but the fact that the electrons fail to contribute their "fair share" to the heat capacities also is inexplicable in the classical theory. In the present section we shall continue to ignore the contribution of the electrons to the heat capacity. The justification of this omission, which in fact is usually quite legitimate, can only be given after we have examined the behavior of the "electron gas" in solids. The study of such an electron gas will be taken up in Section 15. In the present section we shall study the internal energy and the heat capacity that arise from the motion of the atoms in the lattice of solids.

It will be recalled (see Chapter 2) that one model of a solid is to view it as a collection of $3N$ harmonic oscillators, or waves (if there are N atoms). The energy of each oscillator is, according to quantum mechanics, $\epsilon_i = (n_i + \frac{1}{2})h\nu_i$, where $n_i = 0, 1, 2, \ldots$ and ν_i characterizes the oscillator. The total energy of the solid is then

$$E = \sum_{i=1}^{3N} (n_i + \tfrac{1}{2})h\nu_i$$

The partition function for this system is therefore

$$Z = \sum_{\text{states}} e^{-\beta E} = \sum_{n_1=0}^{\infty} \cdots \sum_{n_{3N}=0}^{\infty} e^{-\beta(\epsilon_1+\epsilon_2+\ldots+\epsilon_{3N})}$$

$$= \sum_{n_1=0}^{\infty} e^{-\beta[n_1+(1/2)]h\nu_1} \sum_{n_2=0}^{\infty} e^{-\beta[n_2+(1/2)]h\nu_2} \ldots \sum_{n_{3N}=0}^{\infty} e^{-\beta[n_{3N}+(1/2)]h\nu_{3N}}$$

Using the fact that $\sum_{n=0}^{\infty} x^n = 1/(1-x)$, we obtain

$$Z = \left(\frac{e^{-\beta h\nu_1/2}}{1-e^{-\beta h\nu_1}}\right)\left(\frac{e^{-\beta h\nu_2/2}}{1-e^{-\beta h\nu_2}}\right)\cdots \equiv \prod_{i=1}^{3N}\left(\frac{e^{-\beta h\nu_i/2}}{1-e^{-\beta h\nu_i}}\right)$$

This partition function is a special case of the general result we obtained in Section 8 for distinguishable particles (see problems). Taking the logarithm of Z, we obtain

$$\ln Z = \sum_{i=1}^{3N} -\frac{\beta h\nu_i}{2} - \ln(1 - e^{-\beta h\nu_i}) \tag{10.1}$$

It remains to assign values to the frequencies ν_i. In Chapter 2 we discussed the two most common methods for assigning values to the ν_i. In the Einstein model we take all ν_i equal to a common value ν (the value of which depends on the solid in question). In this case

$$\ln Z = -\frac{3N\beta h\nu}{2} - 3N \ln(1 - e^{-\beta h\nu}) \qquad \text{(Einstein model)} \tag{10.2}$$

The second model, which is more accurate, is due to Debye. In this model we take the frequencies ν_i to range continuously over the values $\nu_m \geq \nu \geq 0$, with the number of waves that have frequencies between ν and $\nu + d\nu$ given by

$$dN(\nu) = \left(\frac{12\pi V}{C^3}\right)\nu^2\,d\nu \tag{10.3}$$

The constant C is related to the longitudinal and transverse sound velocities of the waves (C_ℓ and C_t) by $3C^{-3} = 2C_t^{-3} + C_\ell^{-3}$. The maximum frequency ν_m is fixed by the requirement that there are $3N$ waves in the solid, so

$$3N = \int_0^\infty dN(\nu) = \frac{4\pi V \nu_m^3}{C^3} \tag{10.4}$$

In terms of ν_m we can then write (10.3) in the form

$$dN(\nu) = \left(\frac{9N}{\nu_m^3}\right)\nu^2\,d\nu \tag{10.5}$$

Thus, in the Debye model, the summation in Equation (10.1) is replaced by an integral

$$\ln Z = \int_0^{\nu_m} dN(\nu)\left[-\frac{\beta h\nu}{2} - \ln\left(1 - e^{-\beta h\nu}\right)\right] \qquad \text{(Debye model)} \tag{10.6}$$

Obviously the Debye model is more complicated than the Einstein model — but it also is a more accurate description of a solid.

To see the different predictions of these two models, let us first obtain their respective expressions for the internal energy. According to the Einstein model, Equation (10.2), we would have

$$U = -\frac{\partial \ln Z}{\partial \beta} = \frac{3}{2}Nh\nu + \frac{3Nh\nu}{e^{\beta h\nu} - 1} \qquad \text{(Einstein model)} \tag{10.7}$$

The first term is just a constant (independent of T) and represents the contribution of the zero-point energy of all the oscillators. It has no physical (i.e., measurable) significance. The heat capacity of the solid is obtained from the second term in (10.7), namely

$$C_V = \frac{\partial U}{\partial T} = -3Nh\nu\,\frac{h\nu e^{\beta h\nu}}{(e^{\beta h\nu} - 1)^2}\frac{d\beta}{dT}$$

Using a little algebra, this can be written in the form

$$C_V = 3Nk\,\frac{(\beta h\nu)^2}{(e^{\beta h\nu/2} - e^{-\beta h\nu/2})^2}$$

or

$$\boxed{C_V = 3Nk\,\frac{\left(\dfrac{\beta h\nu}{2}\right)^2}{\left[\sinh\left(\dfrac{\beta h\nu}{2}\right)\right]^2} \qquad \text{(Einstein model)}} \tag{10.8}$$

This is exactly the same expression we obtained for $(C_V)_{\text{vib}}$ in a nondegenerate gas [Equation (9.19)] *except* that we now have a factor $3N$ rather than N. This factor of three comes from the fact that the atoms in a solid are three-dimensional harmonic oscillators, whereas in a diatomic molecule they oscillate only along one direction (the connecting axis between the atoms). Because of this similarity, C_V in (10.8) is also shown in Figure 9.2 provided $(C_V/Nk)_{\text{vib}}$ is replaced by $C_V/3Nk$ in the figure.

Now let us look at the Debye model, Equation (10.6). We have

$$U = -\frac{\partial \ln Z}{\partial \beta} = \int_0^{\nu_m} dN(\nu)\left[\frac{1}{2}h\nu + \frac{h\nu e^{-\beta h\nu}}{(1 - e^{-\beta h\nu})}\right]$$

Again the first term in the integrand is a constant resulting from the zero-point energy. The heat capacity is now given by

$$\begin{aligned} C_V = \frac{\partial U}{\partial T} &= \int_0^{\nu_m} dN(\nu)\,\frac{-(h\nu)^2 e^{\beta h\nu}}{(e^{\beta h\nu} - 1)^2}\frac{d\beta}{dT} \\ &= k\int_0^{\nu_m} \frac{(\beta h\nu)^2\, dN(\nu)}{(e^{\beta h\nu/2} - e^{-\beta h\nu/2})^2} \end{aligned} \tag{10.9}$$

To put this in a simpler form, we first use (10.5) to write

$$dN(\nu) = \left(\frac{9N}{\nu_m^3}\right)\nu^2\, d\nu = 9N(\beta h\nu_m)^{-3}(\beta h\nu)^2\, d(\beta h\nu)$$

and then set $u = \beta h\nu$. In terms of this variable, Equation (10.9) can be written as

$$C_V = 9Nk(\beta h\nu_m)^{-3}\int_0^{\beta h\nu_m} \frac{u^4\, du}{(e^{u/2} - e^{-u/2})^2} \tag{10.10}$$

At this point it is useful to define a **Debye temperature**

$$\Theta_D = \frac{h\nu_m}{k} \tag{10.11}$$

which is similar to the characteristic vibrational temperature of molecules, except that Θ_D is in terms of the maximum frequency ν_m. In terms of Θ_D (10.10) can be written

$$\boxed{C_V = 3NkD\left(\frac{\Theta_D}{T}\right) \qquad \text{(Debye model)}} \tag{10.12}$$

where $D(x)$ is known as the **Debye function**

$$D(x) = \frac{3}{x^3}\int_0^x \frac{u^4 e^u}{(e^u - 1)^2}\, du = \frac{12}{x^3}\int_0^x \frac{u^3\, du}{e^u - 1} - \frac{3x}{e^x - 1} \tag{10.13}$$

This function cannot be expressed in terms of more common functions, but it has been tabulated as a function of x [see, for example, M. Abramowitz and I. A. Stern, eds., *Handbook of Mathematical Functions* (New York:

Dover Publications, Inc., 1965), p. 998]. Some of its values are given in the problems. The most important features of $D(x)$ are the limiting values

$$D(x) = \begin{cases} 1 & \text{as } x \to 0 \\ \dfrac{4\pi^4}{5x^3} & \text{as } x \to \infty \end{cases} \tag{10.14}$$

Using these, together with (10.12), we find that the heat capacity of a solid should have the limiting values

$$C_V \simeq \begin{cases} 3Nk & (T \gg \Theta_D) \\ \dfrac{12\pi^4}{5}\left(\dfrac{T}{\Theta_D}\right)^3 Nk & (T \ll \Theta_D) \end{cases} \qquad \text{(Debye model)} \tag{10.15}$$

We see that, at high temperatures, we recover the equipartition result of $C_V = 3Nk$ (corresponding to $3N$ harmonic oscillators, each with an average energy of kT, so that $U = 3NkT$). This value for C_V is also known as the Dulong-Petit law. The same value for C_V is predicted by the Einstein model, Equation (10.8).

As we would expect, the quantum-mechanical models differ from the classical results only at "low" temperatures. Both the Einstein and Debye models predict that the heat capacity of a solid should go to zero as T goes to zero. In fact, from (10.15) and (10.8) we find that they predict

$$\lim_{T\to 0} C_V = \begin{cases} \dfrac{12\pi^4}{5}\left(\dfrac{T}{\Theta_D}\right)^3 Nk & \text{(Debye model)} \\ 3Nk\left(\dfrac{\Theta_E}{T}\right)^2 e^{-\Theta_E/T} & \text{(Einstein model)} \end{cases} \tag{10.16}$$

where $\Theta_E = h\nu/k$ is the "Einstein temperature." These limiting forms for C_V are entirely different functions of the temperature. The Einstein model predicts much lower values of C_V as T goes to zero than does the Debye model. The more realistic Debye model predicts that C_V should go to zero proportional to T^3 — the so-called **Debye's T^3 law.** It has been found experimentally that Debye's T^3 law does accurately describe the temperature dependence of C_V for many solids at low temperatures. This is illustrated schematically in Figure 10.1, where the points represent typical experimental values of C_V. We can fit these points to the Debye curve by making a suitable selection for the value of Θ_D. Since the scattering of the experimental data is greater at higher temperatures, Θ_D is usually picked to yield agreement with the experiments in the low-temperature region, where Debye's T^3 law applies. The important thing to note is that these points cannot be fitted to the Einstein model for *any* value of Θ_E (see problems). By way of comparison, the Einstein curve, Equation (10.8), is also shown in Figure 10.1 (where we have set $\Theta_E = \Theta_D$).

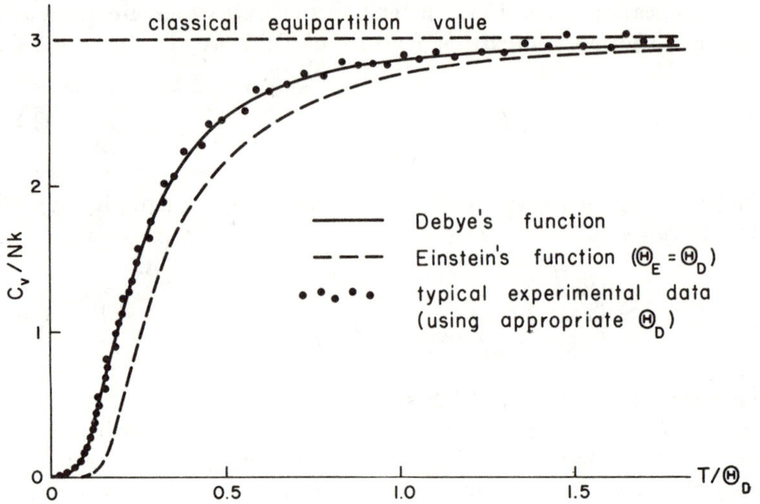

Figure 10.1

Typical values of Θ_D, which have been obtained by this method, are listed in Table 10.1. The solids are listed in the order of increasing Θ_D, ending with the abnormally high value for diamond (most values of Θ_D are below 500°K). The large difference between the heat capacities of lead ($\Theta_D = 88$°K) and diamond ($\Theta_D = 1860$) is very striking (see problems). The fact that such differences in C_V can be represented by one parameter, Θ_D, represents a considerable triumph of the Debye model.

Table 10.1

Solid	Θ_D (°K)	*Solid*	Θ_D (°K)
Lead	88	Copper	315
Potassium	99	Tungsten	315
Sodium	160	Aluminum	390
Silver	220	Iron	430
Zinc	240	Diamond	1860

Although very successful in many respects, the Debye model also has limitations. The fact that some solids cannot be accurately described by one parameter Θ_D should not be especially surprising, considering the variety of complex forces and lattice structures. If Equation (10.12) is used to *define* "the" Debye temperature, then it is found that for some solids Θ_D is not constant but varies with the temperature. An example of this is given in the problems. The Debye model also fails to explain the observed heat capacity of solids at very low temperatures (near 1°K). At these very low temperatures the free electrons in the solid make a significant contribution to C_V (see problems). It was mentioned at the beginning of this section that these contributions can only be explained after we have discussed the Fermi gas (Section 15).

PROBLEMS

1. Refer to the general partition function (8.4) for N distinguishable "particles," and use this result to obtain the present partition function

$$Z = \prod_{i=1}^{3N} \left(\frac{e^{-\beta h\nu_i/2}}{1 - e^{-\beta h\nu_i}} \right)$$

What are the distinguishable particles in the present case, and how many of them are there?

2. To see the significant difference between the Einstein and Debye models, we must plot C_V at low temperatures. Use the expressions (10.16) to plot $C_V/3Nk$ for $0.05 \leq T/\Theta_D \leq 0.1$. In order to fix the value of Θ_E/Θ_D, make the reasonable assumption that the frequency ν in the Einstein model equals one-half the ν_m in the Debye model. Note the difference in the slopes of the resulting curves. To simplify the computation, note that $(1/4x^2)\, e^{-1/2x} = (e^{-1/4x}/2x)^2$.

3. Some values of the Debye function, Equation (10.13), are given in the following table:

x	$D(x)$	x	$D(x)$
0	1.000	4.0	0.503
0.5	0.988	4.5	0.432
1.0	0.952	5.0	0.369
1.5	0.896	6.0	0.266
2.0	0.825	7.0	0.190
2.5	0.745	8.0	0.138
3.0	0.663	10.0	0.076
3.5	0.581	20.0	0.0095

To illustrate the large difference between C_V for lead and diamond, use this table to plot $C_V/3Nk$ as a function of T ($0 \leq T \leq 500$°K) for both of these solids.

4. The experimental values of C_V for cadmium at low temperatures are

T (°K)	1	2	10	20	25	30	40	50	100
C_V (cal/mole-°K)	0.000216	0.00089	0.216	1.24	1.78	2.31	3.15	3.80	5.28

(a) Does C_V obey Debye's T^3 law for $T \leq 2$°K?

(b) Use Equation (10.12) and the table in problem 3 to determine Θ_D for each value of $T \geq 10$°K. Plot $\Theta_D(T)$.

5. At low temperatures, where the Debye T^3 law applies, most substances are in a solid state. An interesting exception is the case of helium, which apparently remains in a liquid state down to 0°K if the pressure is not above 25 atmospheres. A liquid readily propagates longitudinal waves but, unlike a solid, it does not sustain transverse waves. This suggests that the heat capacity of a liquid at low temperatures might be described by a modified Debye theory that does not involve transverse waves. This amounts to replacing C^3 in (10.4) by $3C_l^3$, where C_l is the longitudinal sound velocity.

(a) Using the observed density (0.145 gm/cm³) and sound velocity ($C_l = 2.39 \times 10^4$ cm/sec) of He^4, determine the Debye temperature.

(b) The observed specific heat capacity below 0.6°K is $0.0235T^3$ joules/gm-deg. Using the result of part (a), compare this with the Debye T^3 law.

11. SIMPLE DIELECTRIC SYSTEMS

In this section we shall consider the behavior of systems that contain molecules with a dipole moment when subjected to an externally applied electric field $\boldsymbol{\mathcal{E}}$. A simple model of such systems was discussed in Chapter 2. The potential energy of a molecule, due to the applied electric field, is

$$\Phi = -\boldsymbol{\mathcal{E}}\cdot\mathbf{p} - \tfrac{1}{2}\alpha\mathcal{E}^2 \tag{11.1}$$

where $\mathbf{p}$ is the *permanent* dipole moment of the molecule and α is its polarizability. Thus $\mathbf{p}$ represents the natural asymmetry of the charges in the molecule, whereas $\alpha\boldsymbol{\mathcal{E}}$ is a dipole moment that is induced in the molecule by the applied field. In the simplest systems (simple dielectrics) the dipole moments of the various molecules *do not interact* with one another. This usually means that the dielectric is in a gaseous state (or dilute solution — see problems), so that the molecules are not too close together. In this case the total energy of the system, *due to the electric field*, is simply the sum of the potential energies (11.1), or

$$E = E_0 - \tfrac{1}{2}N\alpha\mathcal{E}^2 - \sum_{i=1}^{N} \boldsymbol{\mathcal{E}}\cdot\mathbf{p}_i \tag{11.2}$$

The term E_0 represents the energy of the system when the electric field is zero. We need not specify the form of this energy because we shall be concerned primarily with the effects produced by the other terms in (11.2). Thus E_0 contains all the energy factors of the molecules that do not relate to the electric dipoles. Finally, we should again note that the energy (11.2) is equally correct for the quantum-mechanical theory, essentially because the orientation of electric dipoles is not quantized (see Chapter 2, Section 5).

Because the energies in (11.2) are additive, the partition function can be written in the form of a product of terms

$$Z = Z_0(T, V) \int_0^{2\pi} d\phi_1 \int_0^{\pi} d\theta_1 \sin\theta_1 \cdots \int_0^{2\pi} d\phi_N \int_0^{\pi} d\theta_N \sin\theta_N \exp\left[\beta\left(\tfrac{1}{2}N\alpha\mathcal{E}^2 + \sum \boldsymbol{\mathcal{E}}\cdot\mathbf{p}_i\right)\right] \qquad (11.3)$$

The term $Z_0(T, V)$ represents the partition function when $\boldsymbol{\mathcal{E}} = 0$, and the remaining terms are the contributions from all the possible orientations of the N dipoles, $\mathbf{p}_i$ (see Figure 11.1). Since the integrands do not depend on

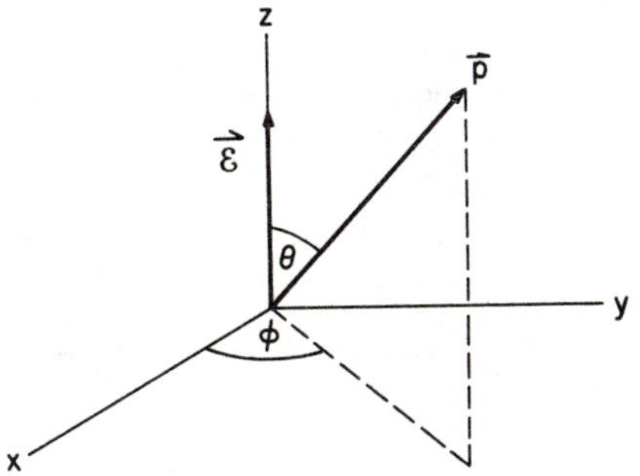

Figure 11.1

the angles ϕ_i, these integrals simply yield a factor of $(2\pi)^N$. Therefore (11.3) equals

$$Z = Z_0(T, V)(2\pi)^N e^{\beta N\alpha\mathcal{E}^2/2}\left(\int_0^{\pi} e^{\beta\mathcal{E}p\cos\theta_1}\sin\theta_1\, d\theta_1\right) \cdots \left(\int_0^{\pi} e^{\beta\mathcal{E}p\cos\theta_N}\sin\theta_N\, d\theta_N\right)$$

Furthermore, since all these integrals have the same value, we readily obtain

$$Z = Z_0(T, V)(2\pi)^N e^{\beta N\alpha\mathcal{E}^2/2}\left(-\int_1^{-1} e^{\beta\mathcal{E}py}\, dy\right)^N$$

$$= Z_0(T, V)\left[\frac{2\pi e^{\beta\alpha\mathcal{E}^2/2}(e^{\beta\mathcal{E}p} - e^{-\beta\mathcal{E}p})}{\beta\mathcal{E}p}\right]^N$$

or finally

$$\boxed{Z(T, V, \mathcal{E}) = Z_0(T, V)\left(\frac{4\pi e^{\beta\alpha\mathcal{E}^2/2}\sinh(\beta\mathcal{E}p)}{\beta\mathcal{E}p}\right)^N} \qquad (11.4)$$

where we have introduced the hyperbolic sine function $\sinh(x) = \frac{1}{2}(e^x - e^{-x})$. Using this partition function, we can now predict the thermodynamic properties of this system.

We note first that some of the thermodynamic properties are unchanged by the interaction of the electric field with the dipoles. For example, the pressure of the system is given, as usual, by the expression

$$\text{pressure} = kT \frac{\partial \ln Z}{\partial V} = kT \frac{\partial \ln Z_0}{\partial V}$$

The last equality is due to the fact that the second term in (11.4) (containing $\mathcal{E}$) does not depend on V. Hence the pressure is given by Z_0 alone and therefore is unaffected by the electric field or the dipole moments. This is a result of our assumption that the electric dipoles do not interact with each other (interaction between electric dipoles, for example, produce the van der Waals forces in imperfect gases). The internal energy, on the other hand, is affected by the applied electric field. This is to be expected, since the average energy of the molecules depends on the interaction between the electric dipoles and the applied field $\boldsymbol{\mathcal{E}}$. We shall return to this point a little later.

The new thermodynamic property of a dielectric system is the *total average dipole moment of the molecules*, which we shall denote by $\boldsymbol{\mathcal{P}}$. Since the total dipole moment of the system is $N\alpha\boldsymbol{\mathcal{E}} + \sum_{i=1}^{N} \mathbf{p}_i$, its average value is given by

$$\boldsymbol{\mathcal{P}} = Z^{-1} \sum_{\text{ms}} \left(N\alpha\boldsymbol{\mathcal{E}} + \sum_{i=1}^{N} \mathbf{p}_i\right) e^{-\beta E} \tag{11.5}$$

where the summation (or integration for the continuous variables) is over all the microstates (ms) of the system. The polarization of the system is $\boldsymbol{\mathcal{P}}/V$, the average dipole moment per unit volume (see problems). We shall refer to $\boldsymbol{\mathcal{P}}$ as the **total polarization** of the system.

Now, if we use the notation

$$\frac{\partial}{\partial \boldsymbol{\mathcal{E}}} = \frac{\partial}{\partial \mathcal{E}_x}\mathbf{i} + \frac{\partial}{\partial \mathcal{E}_y}\mathbf{j} + \frac{\partial}{\partial \mathcal{E}_z}\mathbf{k} \tag{11.6}$$

we find that, according to (11.2),

$$\frac{\partial E}{\partial \boldsymbol{\mathcal{E}}} = -\left(N\alpha\boldsymbol{\mathcal{E}} + \sum_{i=1}^{N} \mathbf{p}_i\right)$$

which is just minus the total dipole moment; therefore (11.5) can be written in the form

$$\boldsymbol{\mathcal{P}} = Z^{-1}\beta^{-1} \frac{\partial}{\partial \boldsymbol{\mathcal{E}}} \sum_{\text{ms}} e^{-\beta E} = Z^{-1}\beta^{-1} \frac{\partial Z}{\partial \boldsymbol{\mathcal{E}}}$$

or

$$\boxed{\boldsymbol{\mathcal{P}} = \beta^{-1} \frac{\partial \ln Z}{\partial \boldsymbol{\mathcal{E}}}} \tag{11.7}$$

(So, for example, $\mathcal{P}_x = \beta^{-1}\,\partial \ln Z/\partial \mathcal{E}_x$). Equation (11.7) shows that the total polarization is also obtained from the logarithm of the partition function, just like other thermodynamic properties. Using (11.4), we find the logarithm of Z to be

$$\begin{aligned}\ln Z = \ln Z_0(T, V) + \tfrac{1}{2}N\beta\alpha\mathcal{E}^2 - N \ln (\beta\mathcal{E}p)\\ + N \ln [\sinh (\beta\mathcal{E}p)] + N \ln 4\pi\end{aligned} \tag{11.8}$$

which depends only on the magnitude $\mathcal{E} = (\mathcal{E}_x^2 + \mathcal{E}_y^2 + \mathcal{E}_z^2)^{1/2}$ of the electric field. Now it is not difficult to show, using (11.6), that

$$\frac{\partial F(\mathcal{E})}{\partial \boldsymbol{\mathcal{E}}} = \frac{\partial F(\mathcal{E})}{\partial \mathcal{E}}\left(\frac{\boldsymbol{\mathcal{E}}}{\mathcal{E}}\right) \tag{11.9}$$

for any function of $\mathcal{E} = |\boldsymbol{\mathcal{E}}|$ (see problems). Therefore, if we substitute (11.8) into (11.7), we obtain

$$\boldsymbol{\mathcal{P}} = \left[N\alpha\mathcal{E} - \left(\frac{N}{\beta\mathcal{E}}\right) + \frac{Np \cosh (\beta p\mathcal{E})}{\sinh (\beta p\mathcal{E})}\right]\left(\frac{\boldsymbol{\mathcal{E}}}{\mathcal{E}}\right)$$

This shows that the total polarization $\boldsymbol{\mathcal{P}}$ is parallel to the applied electric field $\boldsymbol{\mathcal{E}}$. Therefore, we can henceforth just discuss the magnitude of the total polarization

$$\boxed{\mathcal{P} = N\alpha\mathcal{E} + Np\left[\coth (\beta p\mathcal{E}) - \left(\frac{1}{\beta p\mathcal{E}}\right)\right]} \tag{11.10}$$

The first term in (11.10) $N\alpha\mathcal{E}$ represents the total induced polarization in the dielectric. It will be noted that this is independent of the temperature. The second group of terms in (11.10) represents the average polarization due to the permanent dipole moments of the molecules. These terms do depend on the temperature, and the physical reason is not hard to understand. The applied field $\boldsymbol{\mathcal{E}}$ tries to align the permanent dipoles $\mathbf{p}_i$ in a direction parallel to $\boldsymbol{\mathcal{E}}$. However, because of the thermal motion, the dipoles will tend to oscillate about this direction (see Figure 11.2). Therefore, since the orientation of the various dipoles tends to be random, the total permanent dipole moment $\sum_{i=1}^{N} \mathbf{p}_i$ will be less than Np (the value that would be obtained if all the dipoles were parallel). The term [coth $(\beta p\mathcal{E})$ −

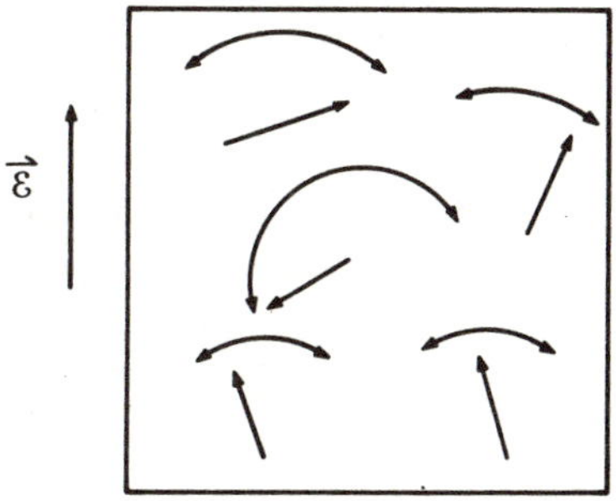

Figure 11.2

$(1/\beta p\mathcal{E})$] in (11.10) represents this reduction in the total polarization produced by the thermal oscillations of the $\mathbf{p}_i$. Presumably this term should become smaller as the temperature increases (increased random orientation). On the other hand, the induced dipole moment of a molecule $\alpha\boldsymbol{\mathcal{E}}$ is *always* parallel to $\boldsymbol{\mathcal{E}}$ and therefore is unaffected by the thermal motion (for it does not oscillate). Consequently the total induced dipole moment is simply $N\alpha\mathcal{E}$ [the first term in (11.10)].

Having seen the qualitative features of (11.10), let us now examine its quantitative behavior. The quantitative properties are determined by the **Langevin function**

$$L(x) = \coth(x) - \left(\frac{1}{x}\right) \tag{11.11}$$

named in honor of P. Langevin, who used this function (1905) to calculate the average magnetic moment of paramagnetic systems (see the next section). The present result for dielectrics was originally obtained by P. Debye in 1912. The behavior of $L(x)$ is shown in Figure 11.3. For small values of

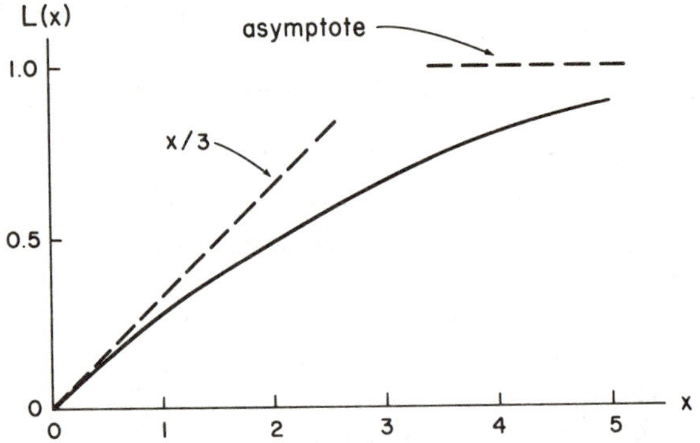

Figure 11.3

x, $L(x) \simeq x/3$, whereas for large values of x, $L(x) \simeq 1$. In terms of the Langevin function the total polarization, Equation (11.10), can be written in the form

$$\boxed{\mathcal{P} = N\alpha\mathcal{E} + NpL(\beta p\mathcal{E})} \tag{11.12}$$

Typical values of p are around 10^{-29} coul-m (see Table 1 in Chapter 2). This number for p corresponds roughly to two charges of $\pm e$ ($e = 1.6 \times 10^{-19}$ coulombs) separated by a distance of 10^{-10} m (= 1 angstrom), which is a typical atomic dimension. At $T = 300°$K, $kT \simeq 4 \times 10^{-21}$ joule, so that for an electric field $\mathcal{E} = 4 \times 10^{3}$ volts/m, the quantity $\beta p\mathcal{E} \simeq 10^{-5}$.

Therefore, at ordinary temperatures, and if $\mathcal{E}$ is not extremely large, the argument of $L(\beta p\mathcal{E})$ is very small. In this case $L(\beta p\mathcal{E}) \simeq \beta p\mathcal{E}/3$ (see Figure 11.3), and Equation (11.12) reduces to

$$\mathcal{P} \simeq N\alpha\mathcal{E} + \left(\frac{Np^2\mathcal{E}}{3kT}\right) \tag{11.13}$$

This is of the form

$$\mathcal{P} = \left(a + \frac{b}{T}\right)\mathcal{E} \tag{11.14}$$

so that if $(\mathcal{P}/\mathcal{E})$ is plotted as a function of $1/T$, one should observe a function such as shown in Figure 11.4. The value of $\mathcal{P}$ as T goes to infinity (that is, $\mathcal{P} = a\mathcal{E}$) is due to the polarizability of the molecules, and the temperature-dependent part is due to the permanent dipole moments of the molecules.

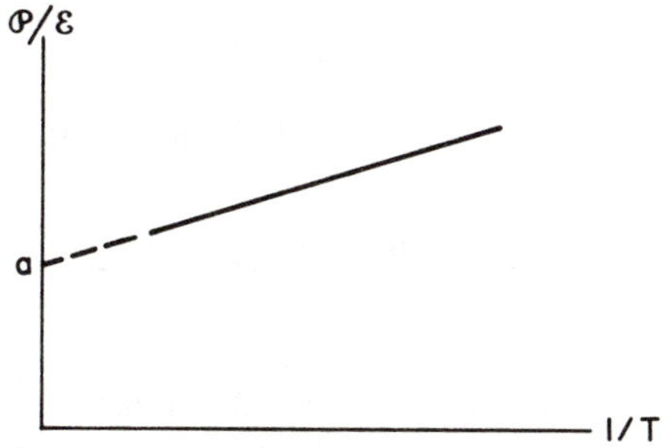

Figure 11.4

By experimentally determining the curve in Figure 11.4, one can deduce the polarizability α and dipole moment p of molecules (see problems).

Finally, let us return to the internal energy and the second law of thermodynamics. The internal energy as given by

$$U = -\frac{\partial \ln Z}{\partial \beta} \tag{11.15}$$

represents the *average total energy* in the system, which obviously is dependent on the electric field $\mathcal{E}$. On the other hand, the "internal energy" that appears in the usual expressions of the second law is *not* the same as (11.15). Since this can cause some confusion, we shall now examine this point. The second law is usually written in the form

$$dU' = T\,dS - p\,dV + \mathcal{E}\,d\mathcal{P} \tag{11.16}$$

where $\mathcal{E}\,d\mathcal{P}$ represents the work done on the system when the total polarization is changed. Of course U' is written without the prime, but we wish to distinguish it from (11.15). To understand the meaning of U′, consider the differential of ln $Z(T, V, \mathcal{E})$:

$$d \ln Z = \frac{\partial \ln Z}{\partial \beta} d\beta + \frac{\partial \ln Z}{\partial V} dV + \frac{\partial \ln Z}{\partial \mathcal{E}} d\mathcal{E}$$

$$= -U\, d\beta + \beta p\, dV + \beta \mathcal{P}\, d\mathcal{E}$$

where we have used (11.15), (11.7), and $p = kT(\partial \ln Z/\partial V)$ (p is the pressure). If we write this in the form

$$d \ln Z = -d(\beta U) + \beta\, dU + \beta p\, dV + \beta d(\mathcal{E}\mathcal{P}) - \beta \mathcal{E}\, d\mathcal{P}$$

and multiply by β^{-1}, then with a little rearranging this becomes

$$d(U + \mathcal{E}\mathcal{P}) = kT\, d(\ln Z + \beta U) - p\, dV + \mathcal{E}\, d\mathcal{P} \tag{11.17}$$

Comparing (11.17) with (11.16), we see that, as usual,

$$dS = kd(\ln Z + \beta U) \tag{11.18}$$

but now

$$U' = U + \mathcal{E}\mathcal{P} \tag{11.19}$$

Using this result, it is not very difficult to show (see problems) that the "internal energy" U' is given by

$$U' = -\frac{\partial \ln Z_0(T, V)}{\partial \beta} + \frac{N\alpha\mathcal{E}^2}{2} \tag{11.20}$$

where Z_0 is the partition function in the absence of an applied field [see (11.4)]. Thus, the internal energy in (11.16) refers to the average energy in the absence of an external electric field, plus the work done by the $\mathcal{E}$ field in making the *induced* dipoles. In other words, U' does not contain any contribution due to the permanent dipole moments of the molecules. If the second law of thermodynamics is written in terms of U, one finds from (11.17) that

$$dU = T\, dS - p\, dV - \mathcal{P}\, d\mathcal{E} \tag{11.21}$$

Thus (11.16) and (11.21) are both correct, but the bookkeeping for the energy differs by the different meanings of "internal energy" in the two cases.

PROBLEMS

1. The polarization of a system is defined as $P = \mathcal{P}/V$ (the average dipole moment per unit volume). The electric susceptibility χ_e (which is dimensionless) is given by

$P = \chi_e\mathcal{E}$	(esu)	$P = \epsilon_0\chi_e\mathcal{E}$	(mks)
		$\epsilon_0 = 8.85 \times 10^{-12}$	(coul/volt-m)

so that $\chi_e(T)$ can be deduced from Equation (11.13). The dielectric constant κ (dimensionless) is related to χ_e by

$$\kappa = 1 + 4\pi\chi_e \quad \text{(esu)} \quad | \quad \kappa = 1 + \chi_e \quad \text{(mks)}$$

and has the same numerical value in both systems of units. Then the force between two charges is given by

$$\frac{q_1q_2}{\kappa r^2} \quad \text{(esu)} \quad \Bigg| \quad \frac{q_1q_2}{4\pi\kappa\epsilon_0 r^2} \quad \text{(mks)}$$

If two charges are in a dielectric, does the force between them increase or decrease if the temperature is increased? Give a brief qualitative (physical) explanation for this effect.

2. Using the fact that $\mathcal{E} = (\mathcal{E}_x^2 + \mathcal{E}_y^2 + \mathcal{E}_z^2)^{1/2}$, prove Equation (11.9), with the help of (11.6).

3. The following table contains experimental values for $\mathcal{P}/\mathcal{E}$ (cm^3) as a function of temperature for one gm-mole of ammonia (NH_3) gas. Using these values, determine the molecular polarizability α and the permanent dipole moment of an ammonia molecule. Compare these values with those given in Table 1, Chapter 2.

T (°K)	292	309	333	387	413	446
$\mathcal{P}/\mathcal{E}$ (cm^3)	13.74	13.13	12.22	10.74	10.15	9.44

4. The electric susceptibility χ_e (in esu system, see Problem 1) of water vapor is found to be

T (°K)	393	423	453	483
Pressure (mm Hg)	563	609	653	697
$\chi_e \times 10^5$	31.8	29.6	27.9	26.2

Assuming that the vapor is a perfect gas, determine the polarizability α (cm^3) and permanent dipole moment (esu-cm). Compare these values with those given in Table 1, Chapter 2.

5. Use Equations (11.8), (11.10), (11.15), and (11.19) to derive Equation (11.20).

6. Use Equations (11.8), (11.15), and (11.18) to obtain an expression for $(S - S_0)/kN$, where S_0 is the entropy for zero electric field. Show that this contribution to the entropy (due to the dipole moments) is independent of the polarizability α. Plot the behavior of this entropy as a function of $x = \beta p\mathcal{E}$ for $5 \geq x \geq 0$. Interpret the qualitative behavior of $S - S_0$ as a function of x in terms of the uncertainty of the dipole orientations. Why is this uncertainty independent of the induced dipole moments? (You may find that Problem 14 of Chapter 1 is of some help.)

12. PARAMAGNETIC SYSTEMS

Systems that contain molecules with a permanent magnetic moment μ_m react to a magnetic field B in much the same way as systems of electric dipoles react to an electric field. The principal difference between these two systems is due to quantum-mechanical effects, which we shall examine in the present section.

The potential energy of a molecule with a permanent magnetic moment $\boldsymbol{\mu}_m$ in a magnetic field $\mathbf{B}$ is $\Phi = -\boldsymbol{\mu}_m \cdot \mathbf{B}$. If the magnetic moments of the molecules do not interact with one another, the total energy of the system is

$$E = E_0 - \sum_{i=1}^{N} \boldsymbol{\mu}_{mi} \cdot \mathbf{B} \tag{12.1}$$

where E_0 is the total energy of the system when $B = 0$. The terms in (12.1) that contain the permanent magnetic moments are referred to as the *paramagnetic* terms. We could also add to (12.1) the diamagnetic energy $-\frac{1}{2}N\alpha_m B^2$, which arises from the induced magnetic moment of molecules $\alpha_m \mathbf{B}$. If this were done, Equation (12.1) would look very similar to equation (11.2) for dielectric systems. However, for simplicity, we shall retain only the paramagnetic terms (paramagnetic systems) in the energy (12.1). In this case (12.1) looks similar to the total energy of a dielectric with zero polarizability ($\alpha = 0$):

$$E = E_0 - \sum_{i=1}^{N} \mathbf{p}_i \cdot \boldsymbol{\mathcal{E}}$$

Indeed, if we could treat the orientation of $\boldsymbol{\mu}_m$ as a continuous variable (similar to the way we were able to treat $\mathbf{p}$), then all the results of Section 11 would also apply here — simply by our making the substitutions

$$\boldsymbol{\mathcal{E}} \to \mathbf{B}, \qquad \mathbf{p} \to \boldsymbol{\mu}_m \tag{12.2}$$

In particular, the **total magnetization** of the system (total average magnetic moment) could be obtained from (11.10):

$$\mathfrak{M} = N\mu_m\left[\coth(\beta\mu_m B) - \left(\frac{1}{\beta\mu_m B}\right)\right] \equiv N\mu_m L(\beta\mu_m B) \tag{12.3}$$

This classical result for the total magnetization was first obtained by P. Langevin in 1905 [prior to Debye's result (11.10) for the dielectric system]. Using the fact that the Langevin function $L(x) \simeq x/3$ for small x, we find that (12.3) reduces to

$$\mathfrak{M} \simeq \frac{\beta N\mu_m^2 B}{3} = \left(\frac{N\mu_m^2}{3k}\right)\left(\frac{B}{T}\right) \tag{12.4}$$

if $\beta\mu_m B \ll 1$ (i.e., weak magnetic fields or high temperatures). This form for the total magnetization is known as **Curie's law.**

$$\mathfrak{M} = C\left(\frac{B}{T}\right) \qquad (C\text{: Curie constant}) \tag{12.5}$$

Thus the classical theory (12.4) gives the value

$$C = \frac{N\mu_m^2}{3k} \tag{12.6}$$

for the Curie constant.

Now let us examine the modifications of these results that arise from quantum-mechanical effects. We noted in Chapter 2, Section 5, that the component of $\boldsymbol{\mu}_m$ along $\mathbf{B}$ can have only certain discrete values. Because of this the potential energy is given by†

$$-\boldsymbol{\mu}_m \cdot \mathbf{B} = -2\mu_B MB \tag{12.7}$$

where $\mu_B = eh/4\pi m_e = 0.927 \times 10^{-23}$ amp-m² (or $eh/4\pi m_e c = 0.927 \times 10^{-20}$ erg/gauss) is the so-called Bohr magneton, and M is restricted to half-integer values between J and $-J$. Examples of such paramagnetic substances, together with their value of J, are listed in Table 12.1. The active

Table 12.1

Substance	*Active Ion*	*J*
Gadolinium sulfate, $Gd_2(SO_4)_3 \cdot 8H_2O$	Gd^{+++}	$\frac{7}{2}$
Ammonium iron alum, $NH_4Fe(SO_4)_2 \cdot 12H_2O$	Fe^{+++}	$\frac{5}{2}$
Potassium chrome alum, $KCr(SO_4)_2 \cdot 12H_2O$	Cr^{+++}	$\frac{3}{2}$

ions, which are the only magnetically active parts in these elaborate molecules, do not interact appreciably because they form only a dilute "solution" among all the other molecules. Therefore, our basic assumption of no magnetic interactions is satisfied for such solids.

Now, since the energy (12.7) has only discrete values, the classical partition function

$$Z = Z_0(T, V)\left(\int_0^{2\pi} d\phi_1 \int_0^{\pi} d\theta_1 \sin\theta_1\, e^{\beta\mu_m B\cos\theta_1}\right) \cdots \left(\int_0^{2\pi} d\phi_N \int_0^{\pi} d\theta_N \sin\theta_N\, e^{\beta\mu_m B\cos\theta_N}\right)$$

†We shall consider only the case of the more important paramagnetic molecules, whose magnetic moment arises from electron spins (an intrinsic angular momentum of the electron). In this case J is the quantum number for the total electron spin of a molecule.

is replaced by

$$Z = Z_0(T, V)\left(\sum_{M_1=-J}^{J} e^{\beta 2\mu_B M_1 B}\right)\cdots\left(\sum_{M_N=-J}^{J} e^{\beta 2\mu_B M_N B}\right)$$

or, since all these series are equal,

$$Z = Z_0(T, V)\left(\sum_{M=-J}^{J} e^{\beta 2\mu_B M B}\right)^N \tag{12.8}$$

We can express this series in the partition function in terms of elementary functions by using the following results. Let $x = \beta 2\mu_B B$, and note that

$$\sum_{M=-J}^{J} e^{Mx} = e^{-Jx}(1 + e^x + e^{2x} + \cdots + e^{2Jx}) \tag{12.9}$$

Now the terms in the parentheses look similar to

$$\frac{1}{1 - e^x} = 1 + e^x + e^{2x} + \cdots + e^{2Jx} + e^{(2J+1)x} + e^{(2J+2)x} + \cdots \tag{12.10}$$

except that the last series contains an infinite number of terms. To obtain the parenthetical series in (12.9), we must subtract from (12.10) the infinite number of terms

$$e^{(2J+1)x} + e^{(2J+2)x} + \cdots = \frac{e^{(2J+1)x}}{1 - e^x} \tag{12.11}$$

The last equality comes from multiplying (12.10) by $e^{(2J+1)x}$. Therefore, if we subtract (12.11) from (12.10), we obtain the parenthetical series in (12.9), so (12.9) becomes

$$\sum_{M=-J}^{J} e^{Mx} = \frac{e^{-Jx}(1 - e^{(2J+1)x})}{1 - e^x}$$

To make this symmetric, multiply the numerator and denominator by $(-e^{-x/2})$, which yields

$$\frac{e^{[J+(1/2)]x} - e^{-[J+(1/2)]x}}{e^{x/2} - e^{-x/2}} = \frac{\sinh [(J + \frac{1}{2})x]}{\sinh (x/2)}$$

Finally, let $x = 2y = 2\beta\mu_B B$, so that

$$\sum_{M=-J}^{J} e^{2My} = \frac{\sinh [(2J + 1)y]}{\sinh (y)}$$

If we use this result in the partition function (12.8), then the logarithm of Z equals

$$\boxed{\ln Z = \ln Z_0(T, V) + N \ln \left\{\frac{\sinh [(2J + 1)y]}{\sinh y}\right\}} \qquad (y = \beta\mu_B B) \tag{12.12}$$

Using this expression, we can obtain the total magnetization, $\mathfrak{M}$ (parallel to B), from the relationship

$$\mathfrak{M} = \beta^{-1} \frac{\partial \ln Z}{\partial B} \tag{12.13}$$

This is analogous to the equation

$$\mathcal{P} = \beta^{-1} \frac{\partial \ln Z}{\partial \mathcal{E}} \tag{11.7}$$

for dielectric systems. Using (12.12) in (12.13), together with the fact that $\partial \ln Z/\partial B = \beta\mu_B(\partial \ln Z/\partial y)$, we obtain

$$\begin{aligned}\mathfrak{M} &= N\mu_B \frac{\partial}{\partial y} \ln \left\{\frac{\sin\ [(2J+1)y]}{\sinh\ (y)}\right\} \\ &= N\mu_B \frac{\sinh\ (y)}{\sinh\ [(2J+1)y]} \cdot \\ &\qquad \left\{\frac{(2J+1)\cosh\ [(2J+1)y]}{\sinh\ (y)} - \frac{\cosh\ (y)\sinh\ [(2J+1)y]}{[\sinh\ (y)]^2}\right\}\end{aligned}$$

which immediately yields

$$\boxed{\mathfrak{M} = N\mu_B\{(2J+1)\coth\ [(2J+1)y] - \coth\ (y)\}} \qquad (y = \beta\mu_B B) \tag{12.14}$$

This expression for the total magnetization, which was first obtained by L. Brillouin, can be compared with the classical (Langevin) expression given in Equation (12.3). To make the comparison complete, we must use the fact that the magnetic moment of a molecule μ_m is related to the Bohr magneton μ_B by the quantum-mechanical expression†

$$\mu_m = 2\mu_B[J(J+1)]^{1/2} \tag{12.15}$$

The odd-looking term $[J(J+1)]^{1/2}$ implies that $\boldsymbol{\mu}_m$ can never be parallel to **B**, because its maximum component along **B** is $2\mu_B J$. This is illustrated schematically in Figure 12.1 for the case $J = \frac{3}{2}$. Since quantum mechanics does not allow $\boldsymbol{\mu}_m$ to be parallel to **B**, whereas classical mechanics does allow this, we expect the classical and quantum values of $\mathfrak{M}$ to differ when (B/T) is large — for when (B/T) is large, $\boldsymbol{\mu}_m$ will tend to be as nearly parallel to **B** *as possible.* Thus the largest

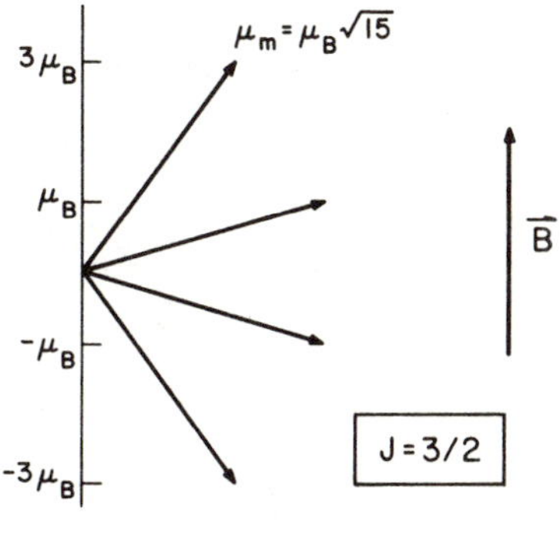

Figure 12.1

†Note again the previous footnote.

classical value of $\mathfrak{M}$ is $N\mu_m$, whereas the largest quantum-mechanical value must be $N2\mu_B J$ (which is less than the classical value). This is sometimes referred to as the saturation value for $\mathfrak{M}$.

To show that these ideas are correct, first consider the limit of weak magnetic field, or high temperature, so that $y = \mu_B B/kT \ll 1$. If we use the approximation

$$\coth(y) \simeq \frac{1}{y} + \frac{y}{3} \qquad (y \ll 1)$$

then Equation (12.14) reduces to

$$\mathfrak{M} \simeq N\mu_B[\tfrac{1}{3}(2J+1)^2 y - \tfrac{1}{3}y]$$

This is of the form of Curie's law

$$\mathfrak{M} = \frac{4}{3}\beta NJ(J+1)\mu_B^2 B = \left(\frac{4NJ(J+1)\mu_B^2}{3k}\right)\left(\frac{B}{T}\right) \qquad (12.16)$$

and it is identical with the classical result (12.4) if μ_m is given by (12.15). In the opposite limit, when $y = \mu_B B/kT \gg 1$, $\coth(y) \simeq 1$, and (12.14) reduces to $\mathfrak{M} = N2\mu_B J$ — in agreement with our previous discussion. The Langevin expression (12.3) yields the larger limiting values of $\mathfrak{M} = N\mu_m = N2\mu_B[J(J+1)]^{1/2}$ when $\beta\mu_m B \gg 1$. Apparently these two results for $\mathfrak{M}$ become nearly equal if J is very large (why? — see problems). The Brillouin expression (12.14) and the Langevin result (12.3) for intermediate values of $y = \beta\mu_B B$ are illustrated in Figure 12.2, for the case $J = \frac{3}{2}$. This shows how the classical and quantum expressions begin to diverge as $\beta\mu_B B$ becomes large. The agreement between experimental measurements for the three

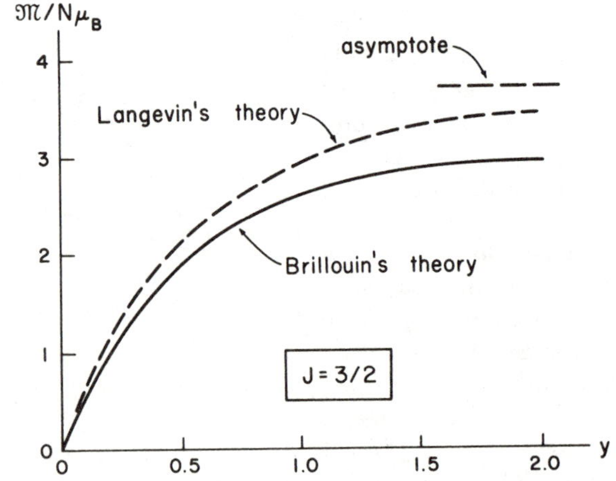

Figure 12.2

paramagnetic substances in Table 12.1 and the Brillouin result (12.14) is illustrated in Figure 12.3. The agreement in the limit of saturated magnetization ($\mu_B B/kT \gg 1$) clearly indicates the correctness of the quantum-mechanical result.

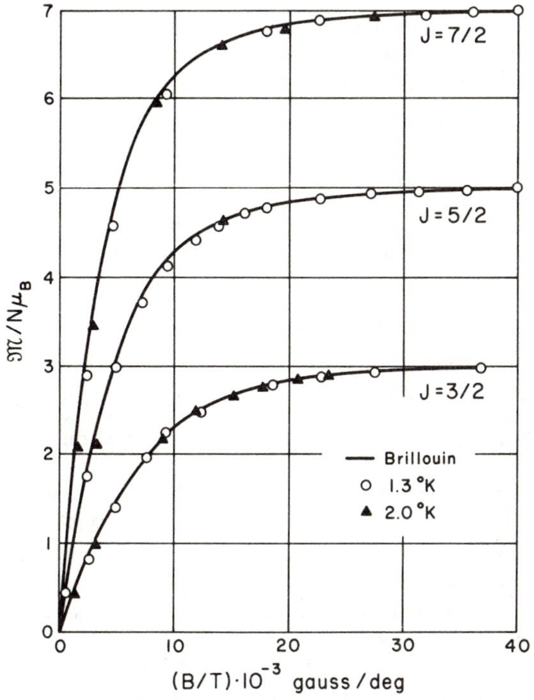

Figure 12.3

PROBLEMS

1. A solid contains noninteracting paramagnetic atoms with a spin $J = \frac{1}{2}$. In this case the possible energy of an atom is $\pm\mu_B B$ [Equation (12.7)]. If $B = 40{,}000$ gauss (4 webers/m²), at what temperature are 75 per cent of the spins oriented in the direction of **B**? At what temperature are 90 per cent of the spins polarized in the **B** direction?

2. Determine the probability of each of the four spin orientations of a paramagnetic atom, shown in Figure 12.1, if $B = 40{,}000$ gauss and $T = 3°K$.

3. The magnetization M is defined as the total magnetization per unit volume $M = \mathfrak{M}/V$. The dimensionless magnetic susceptibility χ_m is then defined by the relationship

$$M = \chi_m H$$

In the present systems involving weak interactions, it is legitimate to make the approximation

$H \simeq B$ (gauss)	$H \simeq \dfrac{B}{\mu_0}$ (B: weber/m^2) $\mu_0 = 4\pi \times 10^{-7}$ weber/amp-m

Oxygen, which is a paramagnetic gas, has a magnetic susceptibility of $\chi_m = 1.44 \times 10^{-7}$ when $T = 293°K$ and the pressure is one atmosphere. Using Curie's law, determine μ_m/μ_B and thereby deduce the value of J for an O_2 molecule.

4. Show that if the internal energy is given (as usual) by

$$U = -\left(\frac{\partial \ln Z}{\partial \beta}\right)_{V,B}$$

then the proper form for the second law of thermodynamics is

$$dU = T\,dS - p\,dV - \mathfrak{M}\,dB$$

(Use an analysis similar to the one at the end of Section 11.)

5. (a) Using the expression

$$S = k(\ln Z + \beta U) = S_0(T, V) + S_m(y), \qquad y = \beta\mu_B B$$

for the entropy, determine the contribution S_m that is due to the magnetic moments of the molecules.

(b) Determine the values of $S_m(y)$ in the limits $y \to \infty$ and $y \to 0$. Explain the significance of these values in terms of the uncertainty of the component of μ_m along the **B** axis.

6. A rather dramatic difference between the classical and quantum theory of paramagnetism can be illustrated by the following example. Consider a paramagnetic system with $J = \frac{1}{2}$.

(a) Using Equation (12.12), and noting that $\sinh(2y) = 2\cosh(y)\sinh(y)$, determine the contribution of the magnetic moments to the internal energy U_{mag}.

(b) The contribution of the magnetic moments to the heat capacity (at constant B) is $C_B = \partial U_{\text{mag}}/\partial T$. Determine C_B, and show that it has a maximum value at $y = 1.20$. Determine its value as $y \to 0$ and $y \to \infty$.

(c) The corresponding classical (Langevin) result is $C_B = Nk[1 - y^2\,\text{csch}^2(y)]$. How does this differ qualitatively with the result of part (b)?

13. RADIATION

The rather unlikely birth of quantum mechanics came about from Planck's study of the thermal equilibrium between electromagnetic radiation and matter. If there is a cavity (an evacuate region of space) inside an opaque substance that is in thermal equilibrium at a temperature T, the cavity will contain electromagnetic radiation. The radiation is continually absorbed and emitted by the wall of the cavity, so that in the cavity there is some average energy density $\rho(\nu, T)\, d\nu$ in the frequency range ν to $\nu + d\nu$. The problem is to predict the function $\rho(\nu, T)$. This problem would be very difficult if it were not for the important fact that $\rho(\nu, T)$ is the same for all cavities, regardless of their shape or the material that forms the walls. The proof of this fact rests on a very beautiful application of the second law of thermodynamics. For our present purpose we shall use the following statement of the second law:

> It is impossible to transfer a net amount of energy between two bodies that are at the same temperature without performing some work on the system.

To prove that $\rho(\nu, T)$ is a universal function that does not depend on the material surrounding the cavity, we consider two cavities A and B shown in Figure 13.1. Let both cavities be in thermal equilibrium at a temperature T. A small pinhole is drilled into each cavity to allow a small amount of radia-

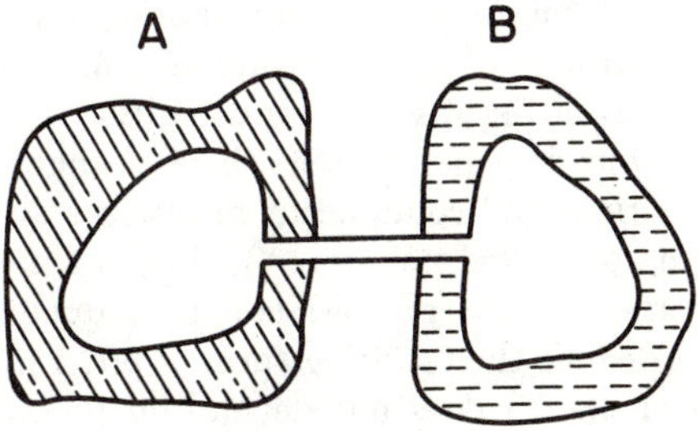

Figure 13.1

tion to be exchanged between the cavities. If, at the end of this exchange of radiation, more energy had passed in one direction than in the other, then one cavity would have an increase in its radiation energy density. When the walls of both A and B returned to equilibrium with the radiation, there would have been a net transfer of energy between materials A and B (one system would have an increase in its temperature, the other a decrease), with no work done on the systems. Since this violates the second law of thermodynamics, we conclude that the energy exchange must be the same in both directions. Obviously this result is independent of the shape of the cavities and the material used in A and B (however, see problems). Furthermore, if we place a filter between the two pinholes that only allows radiation with frequencies in the range between ν and $\nu + d\nu$ to pass between A and B, we see that we could again obtain a violation of the second law unless the equality in energy exchange holds for each value of the frequency. Now the radiation energy that passes through the pinhole is very simply related to the average energy density inside the cavity $\rho(\nu, T)$. The calculation is the same as the one used in Section 3 to determine the average number of particles that pass through a pinhole in a gas container. We showed that this value is $n\bar{v}/4$ per unit area and unit time. In the present case the average number density is replaced by the average energy density $\rho(\nu, T)$ in the frequency range between ν and $\nu + d\nu$. Moreover, the average velocity $\bar{v}$ is replaced by the common velocity of all the radiation, namely the velocity of light c. Thus we have

$$\left(\frac{c}{4}\right)\rho(\nu, T)\,d\nu = \begin{array}{l}\text{average energy that strikes the cavity wall}\\ \text{per unit time and unit area in the frequency}\\ \text{range between } \nu \text{ and } \nu + d\nu\end{array} \tag{13.1}$$

Since we have just shown that the right side of (13.1) must be the same for all cavities in equilibrium, it follows that $\rho(\nu, T)$ is a universal function of ν and T alone.

The function $\rho(\nu, T)$ can be determined experimentally by determining the radiation energy emitted from a cavity for various frequencies. Historically, Planck began with this experimentally determined $\rho(\nu, T)$ and worked backwards to find what laws of nature would lead to such a result (in particular, how the radiation is absorbed and emitted by the walls of the cavity). This forced him to conclude that the exchange of energy between the matter and radiation only could occur in discrete amounts (quanta of radiation, which are now called "photons"). For our present purposes we shall reverse Planck's analysis and assume that quantum mechanics is correct — and then deduce the resulting form of $\rho(\nu, T)$.

Since the form of $\rho(\nu, T)$ does not depend on the composition of the cavity walls, let us assume that the walls consist of harmonic oscillators. According to quantum mechanics (Chapter 2, Section 5) the energy of a harmonic oscillator can only be $\epsilon = (n + \frac{1}{2})h\nu$, where $n = 0, 1, 2, \ldots$; h is Planck's constant; and ν is the characteristic frequency of the oscillator. Assume that when the oscillator (consisting of charged particles) changes its quantum state from n to $n - 1$ (or $n + 1$), it emits (or absorbs) a quanta of radiation (a photon) of frequency ν. The energy of this photon must be $h\nu$ in order to conserve the total energy (radiation plus oscillator). If the oscillator makes the transition from n to $n - 2$, then we can think of it as having emitted two photons (total energy $2h\nu$), and so on, for other transitions. Note that the number of photons in the cavity fluctuates with time, for they can be created and destroyed by the cavity wall. In fact the radiation in the cavity can be viewed as a collection of noninteracting photons, a "photon gas," the number of which is not constant. If the cavity wall contains oscillators with different frequencies ν_i, the photons that they produce will have the energy $h\nu_i$ (and ν_i may be any value). The total energy in the radiation can therefore have only the values

$$E = \epsilon_1 + \epsilon_2 + \epsilon_3 + \cdots, \qquad \text{where } \epsilon_i = n_i h\nu_i$$

Here each n_i can have the values 0, 1, 2, ... (corresponding to 0, 1, 2, ... photons of frequency ν_i). The partition function for the radiation is therefore†

$$Z = \sum e^{-\beta E} = \sum_{n_1=0}^{\infty} \sum_{n_2=0}^{\infty} \cdots e^{-\beta(\epsilon_1+\epsilon_2+\cdots)}$$

$$= \left(\sum_{n_1=0}^{\infty} e^{-\beta n_1 h\nu_1}\right)\left(\sum_{n_2=0}^{\infty} e^{-\beta n_2 h\nu_2}\right)\cdots$$

where $\beta = 1/kT$ refers to the temperature of the cavity walls. Now, using the familiar relationship $\sum_{n=0}^{\infty} x^n = 1/(1-x)$, we have

$$Z = (1 - e^{-\beta h\nu_1})^{-1}\,(1 - e^{-\beta h\nu_2})^{-1} \cdots = \prod_{i=1}^{\infty} (1 - e^{-\beta h\nu_i})^{-1}$$

If we take the logarithm of this, we obtain

$$\ln Z = -\sum_{i=1}^{\infty} \ln\,(1 - e^{-\beta h\nu_i}) = -\sum_{\nu} \ln\,(1 - e^{-\beta h\nu}) \tag{13.2}$$

Now the frequencies of the waves that can exist within a cavity are essentially continuously distributed — provided that the cavity is not too small. If the walls of t he cavity contain some oscillators at all frequencies, then the radiation will also contain all frequencies (see problems). According to the analysis in Chapter 2, Equation (45), the number of waves in a cavity of volume V that have a frequency between ν and $\nu + d\nu$ is $(4\pi V/c^3)\nu^2\,d\nu$. In the present case, since the radiation can be polarized in either of two directions, the total number of waves in the range $d\nu$ of ν is $dN(\nu) = (8\pi V/c^3)\nu^2\,d\nu$, where c is the velocity of light. Then Equation (13.2) becomes

$$\ln Z = -\int dN(\nu)\,\ln\,(1 - e^{-\beta h\nu}) = -\frac{8\pi V}{c^3}\int_0^{\infty} \nu^2 \ln\,(1 - e^{-\beta h\nu})\,d\nu \tag{13.3}$$

†In writing this partition function we assume that the probability of a state does not depend on the total number of photons $(n_1 + n_2 + n_3 + \cdots)$. Since the photon gas is an open system, this means we have assumed that $\mu = 0$ (where μ is the Gibbs free energy per photon — see Chapter 3, Section 7). The justification for this rests, in part, on the agreement between the experimental values for $\rho(\nu, T)$ and the values predicted from the present theory (with $\mu = 0$). When $\mu = 0$, the relationship between Z and the thermodynamic quantities is the same as for a closed system.

From this partition function we can determine $\rho(\nu, T)$ and all the thermodynamic properties of the radiation.

To determine $\rho(\nu, T)$, consider first the internal energy of the radiation, which is given by

$$U = -\frac{\partial \ln Z}{\partial \beta} = \frac{8\pi V}{c^3} \int_0^\infty \frac{h\nu^3 \, d\nu}{e^{\beta h\nu} - 1} = \frac{8\pi^5 (kT)^4 V}{15(hc)^3} \tag{13.4}$$

where we have used (13.3) and the fact that

$$\int_0^\infty \frac{x^3}{e^x - 1} dx = \frac{\pi^4}{15} \tag{13.5}$$

Since U is the average energy of the photons, then by the definition of $\rho(\nu, T)$

$$U = V \int_0^\infty \rho(\nu, T) \, d\nu$$

Comparing this with (13.4), we find that

$$\boxed{\rho(\nu, T) = \frac{8\pi}{c^3} \frac{h\nu^3}{e^{\beta h\nu} - 1}} \tag{13.6}$$

This is one of the forms of the famous Planck distribution for the average energy density. The average energy that can escape through a pinhole in the cavity (per unit time and area) in the range $d\nu$ of ν is, according to (13.1) and (13.6),

$$\boxed{\mathbf{e}(\nu, T) \, d\nu = \frac{2\pi h}{c^2} \frac{\nu^3}{e^{\beta h\nu} - 1} d\nu} \tag{13.7}$$

where $\mathbf{e}(\nu, T)$ is called the **black-body emissivity.** It is this quantity that is measured experimentally and that Planck set out to derive. This function is illustrated in Figure 13.2 for two values of the temperature.

Two features of this figure, which are known from ordinary experiences, should be noted. First, the total radiated energy per unit area and time is larger for larger values of T. Specifically we find

$$\int_0^\infty \mathbf{e}(\nu, T) \, d\nu = \frac{2\pi^5}{15} \frac{(kT)^4}{h^3 c^2} \equiv \sigma T^4 \tag{13.8}$$

where we have again used (13.5). The quantity

$$\sigma = \frac{2\pi^5}{15} \frac{k^4}{h^3 c^2} = 5.669 \times 10^{-5} \text{ erg/cm}^2\text{-sec-deg}^4 \tag{13.9}$$

is known as the **Stefan-Boltzmann constant.** We see that the total radiated energy increases as T^4. This is related to the common experience of feeling more heat (radiation) coming from a hot object. Secondly, as T in-

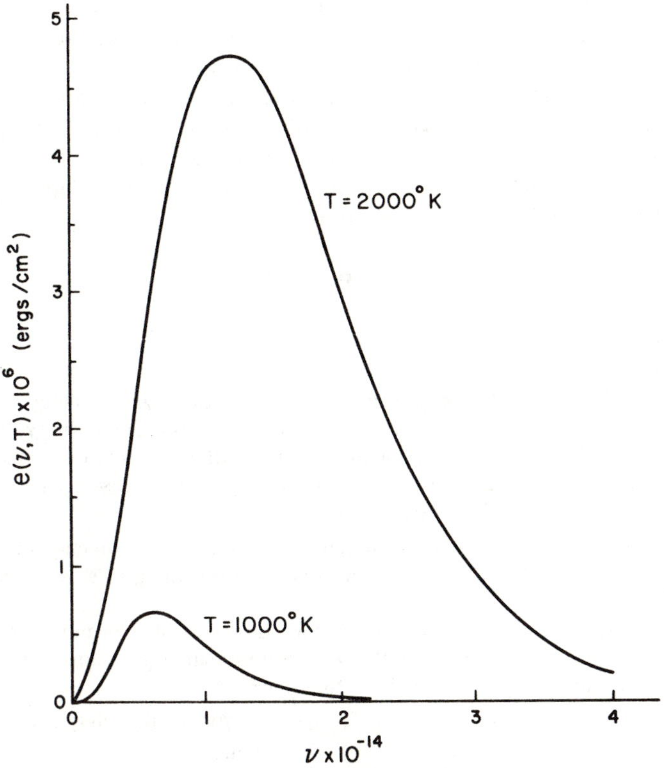

Figure 13.2

creases, the frequency at which the radiated energy is a maximum $\nu_{max} = 2.82kT/h$ also increases. This corresponds to the common experience of seeing the light from a hot body shift from red toward blue (higher frequency) as its temperature is increased.

To obtain the remaining thermodynamic functions, we first note that $\ln Z$, as given by Equation (13.3), equals

$$\ln Z = \frac{8\pi^5 V}{45(\beta hc)^3} \tag{13.10}$$

(see problems). Using the relationship

$$p = kT\left(\frac{\partial \ln Z}{\partial V}\right)_T$$

we find that the radiation pressure is given by

$$p = \frac{8\pi^5}{45(hc)^3\beta^4} = \frac{4\sigma}{3c}T^4 \tag{13.11}$$

where σ is the Stefan-Boltzmann constant, Equation (13.9). Equation (13.11) is therefore the equation of state for the radiation field in a cavity. It will be noticed that the radiation pressure is independent of the volume. This is similar to the behavior of ordinary systems that contain two phases (e.g., liquid-vapor, solid-vapor, or solid-liquid). However, in the present case the pressure increases very rapidly with the temperature (T^4). The remaining thermodynamic functions can be obtained in a similar way from the partition function (13.10) (see problems).

PROBLEMS

1. Using Equation (13.5), obtain Equation (13.10) from Equation (13.3) by integrating by parts.

2. (a) Show how the radiation pressure is related to the internal energy.
(b) Using Equation (13.10), obtain an expression for the entropy, the free energy, and the Gibbs function for the radiation, all in terms of U. Explain why G has this value (using the general expression for G in the case of open systems and the discussion in the present section).

3. (a) At what temperature is the radiation pressure equal to one atmosphere (10^6 dynes/cm^2)? To 10^{-12} mm of mercury (corresponding to a very good vacuum)?
(b) It is estimated that the temperature at the center of the sun is 2×10^7 °K. If it is in equilibrium, determine the radiation pressure and compare it with the estimated "gas" pressure of 4×10^{11} atmospheres.

4. (a) Obtain approximate expressions for $\rho(\nu, T)$ in the limits $h\nu/kT \ll 1$ (Rayleigh-Jeans limit) and $h\nu/kT \gg 1$ (Wiën limit).
(b) Derive the Rayleigh-Jeans limit using the equipartition law, assuming that each radiation mode in the cavity can be thought of as a harmonic oscillator [recall that $dN(\nu) = (8\pi V/c^3)\nu^2\, d\nu$].

5. (a) Using (13.7), show that the energy radiated per unit area and time in the range $d\lambda$ of λ (where $\lambda = c/\nu$ is the wavelength) is

$$\left(\frac{2\pi c^2 h}{\lambda^5}\right)(e^{\beta hc/\lambda} - 1)^{-1}\, d\lambda \equiv \mathbf{e}(\lambda, T)\, d\lambda$$

(b) Show that the wavelength for which $\mathbf{e}(\lambda, T)$ is a maximum is given by

$$\beta hc = 4.965\lambda_{\text{max}}$$

What does $\lambda_{\text{max}}\nu_{\text{max}}/c$ equal?
(c) Solar radiation has a maximum intensity near $\lambda = 5 \times 10^{-5}$ cm. Assuming that the sun's surface is in thermal equilibrium, determine its temperature.

6. (a) Obtain an expression for the average number of photons $\overline{N}(V, T)$ in an equilibrium photon gas, using the fact that

$$\int_0^\infty \frac{x^2\, dx}{e^x - 1} = 2.404$$

[Hint: Use the average energy density and the fact that the energy of a photon is $h\nu$.] Use this result to express the equation of state, Equation (13.11), in terms of p, $\overline{N}$, V, and T. What is the ratio of the number of molecules in a perfect gas to the average number of photons, for the same p, V, and T?
(b) Obtain the heat capacity C_V for a photon gas and, using the result of part (a), express it in terms of $\overline{N}$. Determine the ratio of the specific heat for the radiation c_V (i.e., per photon), to that of a perfect monatomic gas $3k/2$.

7. Recent astronomical observations appear to indicate that the universe is filled with radiation, which may be the adiabatically cooled residuum from a fiery birth of the universe (the "big-bang" theory). Compare the following (indirectly) observed values of the radiation intensity ($\pi^{-1}\mathbf{e}(\nu, T)$ in ergs/cm^2) with those obtained from black-body radiation at $T = 3°$K.

λ (cm)	0.26	3.2	7.5	20.7
Intensity	$(5 \pm 2) \times 10^{-15}$	$(6.5 \pm 1.5) \times 10^{-17}$	$(1.7 \pm 0.8) \times 10^{-17}$	$(1.5 \pm 0.5) \times 10^{-18}$

Are the two sets of values consistent with one another?†

8. In the derivation of $\rho(\nu, T)$ in this section it was assumed that the cavity walls contained some oscillators with every value of the frequency. This is necessary in order for the radiation at any frequency to be able to interact with the wall — and thereby come to thermal equilibrium with the wall. Discuss what would happen to $\rho(\nu, T)$ if the cavity wall did *not* contain oscillators with frequencies $\nu_1 \leq \nu \leq \nu_2$. Would this alter $\rho(\nu, T)$ for frequencies outside this range? Could we obtain a violation of the second law of thermodynamics by coupling this cavity (Figure 13.1) with another cavity containing *all* frequencies?

14. PERFECT GASES (GENERAL THEORY)

The quantum statistics of noninteracting particles was discussed in some detail in Section 8, and it was pointed out that one must distinguish between two possible situations. If the particles of the system are distinguishable, we found that no particularly novel problems arise in evaluating the partition function, except those due to the quantized energy levels. On the other hand, if the particles are indistinguishable, we must introduce the concept of **occupation numbers** n_s defined by

$$n_s = \text{the number of particles in the particle state } s \tag{14.1}$$

†An account of these observations, together with references, can be found in *Physics Today*, **19**, 75 (1966).

A microstate of the system is then specified by the values of all the occupation numbers $n_s = (n_1, n_2, n_3, \ldots)$. Moreover, if we let $\epsilon(s)$ represent the energy of the particle state s, then the total energy of the system of noninteracting particles is

$$E = \sum_{s=1}^{\infty} n_s \epsilon(s) \tag{14.2}$$

For example, if a system has particles with energies

$$\epsilon(1), \quad \epsilon(3), \quad \epsilon(2), \quad \epsilon(1), \quad \epsilon(5)$$

then the microstate is represented by $(2, 1, 1, 0, 1, 0, \ldots)$ and the total energy is $2\epsilon(1) + \epsilon(2) + \epsilon(3) + \epsilon(5)$. The total number of particles in the system is given by

$$N = \sum_{s=1}^{\infty} n_s \tag{14.3}$$

In Section 8 we found that the partition function for a system of indistinguishable particles can be treated fairly easily if the system is *nondegenerate* — that is to say, if the number of particle states with $\epsilon(s) < kT$ is much greater than N. In this case most of the occupation numbers n_s are zero. In the present section we shall consider the case in which the system may be degenerate. As we saw in Section 8, this generally occurs only at very low temperatures or at high densities — in which case quantum-mechanical effects can be expected to be very important.

In order to treat the general situation (i.e., degenerate or nondegenerate) we must take into account one more fact of nature. It has been found that nature apparently consists of only two types of particles, **Bose-Einstein** particles and **Fermi-Dirac** particles. In the case of *Bose-Einstein* particles the possible values of n_s are $0, 1, 2, \ldots, N$. This simply means that any number of particles can be in the same particle state. Some examples of Bose-Einstein particles are H, H_2, and He^4 (or any particle with an integer value of the "spin").† On the other hand, *Fermi-Dirac* particles can have no more than one particle in any particle state, so n_s can be only zero or one. Examples of these particles are electrons, protons, neutrons, H^+, and He^3

†Just as we must assign a mass and a charge to a particle in order to explain its behavior, it has been found that we must also assign a spin to particles. The spin σ of a particle is related to an intrinsic angular momentum — i.e., an angular momentum that is always associated with the particle. For the fundamental particles electron, proton, and neutron it has been found that $\sigma = \frac{1}{2}$. For "compound" particles, such as H_2, the spin is integer or half-integer, depending on whether the compound particle is composed of an even or odd number of these elementary particles.

(or any particle with a half-integer spin). To summarize, we have the two possibilities

$$n_s = \begin{cases} 0, 1, 2, 3, \ldots, N & \text{(Bose-Einstein particles)} \\ 0, 1 & \text{(Fermi-Dirac particles)} \end{cases} \tag{14.4}$$

In the nondegenerate case, treated in Section 8, we did not have to distinguish between these two types of particles, because for most of the states $n_s = 0$. According to (14.4), either of these particles can satisfy this condition, so it was unnecessary to distinguish between them. However, in the degenerate case, when the nubemr of particle states $\epsilon(s) < kT$ is much less than N, the particles will have a tendency to take on the lowest possible value of the energy $\epsilon(s)$. The Fermi-Dirac particles can have only one particle in each particle state, so they cannot all occupy the lowest energy state. The Bose-Einstein particles, on the other hand, can all occupy the lowest energy state. This difference between the two types of particles becomes very important at low temperatures (i.e., in the degenerate case).

In order to predict the thermodynamic properties of the present systems it is very useful to use the partition function for open systems, discussed in Chapter 3. The reason is that the resulting partition function is much simpler to evaluate than the partition function for closed systems (fixed value of N). Moreover, if the fluctuations in the number of particles is very small compared to $\bar{N}$, the thermodynamic properties are the same as for a closed system (with $N = \bar{N}$). Therefore, we begin with the partition function for open systems (the grand canonical partition function)

$$\mathrm{Z} = \sum_{N=0}^{\infty} \sum_{\mathrm{ms}} e^{\beta\mu N - \beta E} \tag{14.5}$$

where the first sum is over the microstates for a system of N particles, and this is followed by a sum over all possible values of N. The parameter μ (which has the dimensions of energy) characterizes the particle reservoir in much the same way as β characterizes the thermal (or energy) reservoir. It was shown in Chapter 3 that μ could be identified with the specific Gibbs function of the system (if that is any help!). Perhaps it would be more enlightening to note that for large negative values of μ, the system is not likely to have a large number of particles (for $e^{\beta\mu N}$ is small if N is large and $\mu \ll 0$). This is analogous to the situation of large (positive) β — in which case the energy is not likely to be large. Unlike β, however, μ can sometimes be either positive or negative, as we shall see shortly.

If we proceed by substituting (14.2) into (14.5), we obtain

$$\mathrm{Z} = \sum_{N=0}^{\infty} \sum_{\{ns\}} \exp\left[\beta\mu N - \beta \sum_{s=1}^{\infty} n_s\epsilon(s)\right] \tag{14.6}$$

where the first sum is over all occupation numbers that satisfy

$$\sum_{s=1}^{\infty} n_s = N \tag{14.3}$$

Equation (14.6) looks rather complicated at first, but we shall now show that $\mathcal{Z}$ can be expressed in a form that does not involve the occupation number $\{n_s\}$. To do this, we shall discuss the case of Bose-Einstein and Fermi-Dirac particles separately.

First of all, consider the case of Fermi-Dirac particles, for which $n_s = 0$ or $n_s = 1$. To make things a little easier, assume there are only three particle states, having energies $\epsilon(1)$, $\epsilon(2)$, and $\epsilon(3)$. The sum on N then must be restricted to $N = 0, 1, 2, 3$, for if N were larger than 3, then two particles would have to be in the same state — and this is not allowed in the Fermi-Dirac case. After we have examined this hypothetical system, it will be easy to generalize our result to an infinite number of states $\epsilon(s)$ and arbitrary values of N. Let us write out $\mathcal{Z}$, as given by (14.6), in detail for this hypothetical system. We have

$$\begin{aligned}\mathcal{Z} = 1 + e^{\beta\mu}(e^{-\beta\epsilon(1)} + e^{-\beta\epsilon(2)} + e^{-\beta\epsilon(3)}) + e^{2\beta\mu}(e^{-\beta[\epsilon(1)+\epsilon(2)]} \\ + e^{-\beta[\epsilon(1)+\epsilon(3)]} + e^{-\beta[\epsilon(2)+\epsilon(3)]} + e^{3\beta\mu}e^{-\beta[\epsilon(1)+\epsilon(2)+\epsilon(3)]}\end{aligned}$$

where the four groups of terms correspond to $N = 0$, 1, 2, and 3, respectively. Now this expression can also be written in the form

$$\mathcal{Z} = (1 + e^{-\beta\mu-\beta\epsilon(1)})(1 + e^{\beta\mu-\beta\epsilon(2)})(1 + e^{\beta\mu-\beta\epsilon(3)}) \equiv \prod_{s=1}^{3} (1 + e^{\beta[\mu-\epsilon(s)]})$$

We may verify this by writing out the terms of this product. For an infinite number of particle states it becomes a product of an infinite number of terms (instead of three), or

$$\mathcal{Z}_{FD} = \prod_{s=1}^{\infty} (1 + e^{\beta[\mu-\epsilon(s)]}) \tag{14.7}$$

This is the desired simplification of (14.6) for the case of Fermi-Dirac particles.

In the case of *Bose-Einstein* particles the situation is different, because each n_s may have any value (up to N). If we consider again the hypothetical system with three particle states $\epsilon(1)$, $\epsilon(2)$, and $\epsilon(3)$, then (14.6) now becomes

$$\begin{aligned}\mathcal{Z} = 1 + e^{\beta\mu}(e^{-\beta\epsilon(1)} + e^{-\beta\epsilon(2)} + e^{-\beta\epsilon(3)}) + e^{2\beta\mu}(\underline{e^{-2\beta\epsilon(1)}} + e^{-\beta[\epsilon(1)+\epsilon(2)]} \\ + \underline{e^{-2\beta\epsilon(2)}} + e^{-\beta[\epsilon(1)+\epsilon(3)]} + \underline{e^{-2\beta\epsilon(3)}} + e^{-\beta[\epsilon(2)+\epsilon(3)]}) + \cdots\end{aligned}$$

The three groups correspond to $N = 0$, 1, and 2, respectively. Now N can go to infinity, so we represent the rest by "· · ·". The underlined terms do not occur in the case of Fermi-Dirac particles, because they correspond to

$n_1 = 2$, $n_2 = 2$, and $n_3 = 2$, respectively. With a little effort (write out a few terms) it is not difficult to see that the above expression is equal to

$$\mathcal{Z} = (1 + e^{\beta[\mu-\epsilon(1)]} + e^{2\beta[\mu-\epsilon(1)]} + \cdots)(1 + e^{\beta[\mu-\epsilon(2)]} + e^{2\beta[\mu-\epsilon(2)]} + \cdots)$$
$$\times (1 + e^{\beta[\mu-\epsilon(3)]} + e^{2\beta[\mu-\epsilon(3)]} + \cdots) = \left(1 + \sum_{n_1=1}^{\infty} e^{n_1\beta[\mu-\epsilon(1)]}\right)$$
$$\times \left(1 + \sum_{n_2=1}^{\infty} e^{n_2\beta[\mu-\epsilon(2)]}\right)\left(1 + \sum_{n_3=1}^{\infty} e^{n_3\beta[\mu-\epsilon(3)]}\right)$$

To reduce this further, we note again that

$$1 + \sum_{n=1}^{\infty} x^n = (1 - x)^{-1} \qquad (\textit{if } x < 1)$$

Now the series that appear in $\mathcal{Z}$ will yield a finite value only if $e^{\beta[\mu-\epsilon]} < 1$,† so by taking $x = e^{\beta[\mu-\epsilon(s)]}$, we obtain the fairly simple expression

$$\mathcal{Z} = (1 - e^{\beta[\mu-\epsilon(1)]})^{-1}(1 - e^{\beta[\mu-\epsilon(2)]})^{-1}(1 - e^{\beta[\mu-\epsilon(3)]})^{-1}$$

If we now return to a real system with an infinite number of energies $\epsilon(s)$ this generalizes to

$$\mathcal{Z}_{BE} = \prod_{s=1}^{\infty} (1 - e^{\beta[\mu-\epsilon(s)]})^{-1} \tag{14.8}$$

Comparing (14.8) and (14.7), we see that the two partition functions can be written

$$\boxed{\mathcal{Z} = \prod_{s=1}^{\infty} (1 \pm e^{\beta[\mu-\epsilon(s)]})^{\pm 1}, \qquad \begin{cases} +: \text{Fermi-Dirac} \\ -: \text{Bose-Einstein} \end{cases}} \tag{14.9}$$

Since all the thermodynamic properties are obtained from the logarithm of $\mathcal{Z}$, rather than $\mathcal{Z}$, we note that

$$\boxed{\ln \mathcal{Z} = \pm \sum_{s=1}^{\infty} \ln (1 \pm e^{\beta[\mu-\epsilon(s)]}), \qquad \begin{cases} +: \text{Fermi-Dirac} \\ -: \text{Bose-Einstein} \end{cases}} \tag{14.10}$$

Having reduced the partition function to a more manageable expression, let us now turn to the question of how the particles are distributed over the various energy levels $\epsilon(s)$. In this classical perfect gas the probable number of particles in the range (d^3r, d^3v) of $(\mathbf{r}, \mathbf{v})$ is represented by $\eta(\mathbf{r}, \mathbf{v})\, d^3r\, d^3v$.

†Therefore, for a Bose-Einstein perfect gas, μ must be less than the lowest value of all $\epsilon(s)$ — which is usually taken to be zero. Then $\mu < 0$. This restriction does not apply in the Fermi-Dirac case (see problems).

In the present quantum case we can ask for the average (or probable) number of particles in a particular particle state (say $s = q$) with an energy $\epsilon(q)$. We shall denote† this quantity by $\bar{n}_q$, so

$$\bar{n}_q = \text{average number of particles in the particle state } q \qquad (14.11)$$

To obtain $\bar{n}_q$ we must take the probability of a particular microstate, multiply it by the value of n_q *in that microstate,* and then sum over all microstates. In other words,

$$\bar{n}_q = \sum_{N=0}^{\infty} \sum_{\text{ms}} n_q P(E, N) = \sum_{N=0}^{\infty} \sum_{\{n_s\}} n_q \mathcal{Z}^{-1} e^{\beta[\mu N - E]}$$

where, again,

$$E = \sum_{s=1}^{\infty} n_s \epsilon(s), \qquad N = \sum_{s=1}^{\infty} n_s$$

Therefore,

$$\begin{aligned}\bar{n}_q &= \mathcal{Z}^{-1} \sum_{N=0}^{\infty} \sum_{\{n_s\}} n_q \exp\left[\beta\mu N - \beta \sum_{s=1}^{\infty} n_s \epsilon(s)\right] \\ &= \mathcal{Z}^{-1} \left(\frac{-1}{\beta}\right) \frac{\partial}{\partial \epsilon(q)} \sum_{N=0}^{\infty} \sum_{\{n_s\}} \exp\left[\beta\mu N - \beta \sum_{s=1}^{\infty} n_s \epsilon(s)\right]\end{aligned}$$

The summations simply yield the partition function (14.6), so

$$\bar{n}_q = -\frac{1}{\beta} \mathcal{Z}^{-1} \frac{\partial \mathcal{Z}}{\partial \epsilon(q)}$$

Therefore,

$$\boxed{\bar{n}_q = -\beta^{-1} \frac{\partial \ln \mathcal{Z}}{\partial \epsilon(q)}} \qquad (14.12)$$

This result is equally valid for the Bose-Einstein or the Fermi-Dirac system. Therefore, if we substitute our previous expression (14.10) for the $\ln \mathcal{Z}$ into (14.12), we obtain $\bar{n}_q$ for either of these systems:

$$\begin{aligned}\bar{n}_q &= -\beta^{-1} \frac{\partial}{\partial \epsilon(q)} \left[\pm \sum_{s=1}^{\infty} \ln\left(1 \pm e^{\beta[\mu - \epsilon(s)]}\right)\right] \\ &= \mp \beta^{-1} \frac{\partial}{\partial \epsilon(q)} \ln\left(1 \pm e^{\beta[\mu - \epsilon(q)]}\right) = \frac{e^{\beta[\mu - \epsilon(q)]}}{1 \pm e^{\beta[\mu - \epsilon(q)]}}\end{aligned}$$

†The subscripts s and q will both be used to denote particle states. The subscript s will usually be used as a summation index, whereas q will refer to some particular particle state.

In other words

$$\bar{n}_q = (e^{-\beta[\mu-\epsilon(q)]} \pm 1)^{-1} \qquad \begin{cases} +\text{: Fermi-Dirac} \\ -\text{: Bose-Einstein} \end{cases} \tag{14.13}$$

We see that the only difference between a Fermi-Dirac system and a Bose-Einstein system is "just" a few plus or minus signs in the partition function and $\bar{n}_q$. These signs, however, signify very different physical properties for the two types of systems. The properties are sufficiently distinct for us to treat these systems separately in the next two sections.

PROBLEMS

1. (a) Use the fact that the total average number of particles is given by (Chapter 3)

$$\bar{N} = kT\left(\frac{\partial \ln \mathcal{Z}}{\partial \mu}\right)_{V,T}$$

and show that this also equals $\sum_s \bar{n}_s$.

(b) What thermodynamic state function is represented by $\sum_s \epsilon(s)\bar{n}_s$? Prove this identification, using the results in Chapter 3.

2. The grand canonical partition function $\mathcal{Z}$ can also be written as

$$\mathcal{Z} = \sum_{N=0}^{\infty} e^{\beta\mu N} Z_N$$

where Z_N is the partition function for a closed system of N particles ($Z_N = \sum_{\text{ms}} e^{-\beta E}$). Use the results of Section 8 to show that for a nondegenerate perfect gas

$$\mathcal{Z} \simeq \exp\{\mathfrak{z}e^{\beta\mu}\}$$

where $\mathfrak{z}$ is the particle partition function.

3. As was mentioned in a footnote, the parameter μ for a Bose-Einstein perfect gas must be less than the lowest value of the particle's energy $\epsilon(s)$ (which we can take to be zero). This is necessary in order for $\mathcal{Z}$ to have a finite value. Now consider Equation (14.5). Why is it possible for $\mu > 0$ in the case of Fermi-Dirac particles? Put another way, why may the terms in the series (14.5) decrease as N increases, even if $\mu > 0$ (thus making it possible for $\mathcal{Z}$ to be finite)? [Hint: Consider Equations (14.2) and (14.4).]

4. (a) According to the results in Chapter 3, the entropy equals (aside from an additive constant)

$$S = k(\ln \mathcal{Z} + \beta U - \beta\mu\bar{N})$$

Using the results of this section, first show that

$$\beta(U - \mu\bar{N}) = -\sum_s \bar{n}_s \ln\left(\frac{\bar{n}_s}{1 \mp \bar{n}_s}\right)$$

where the upper (lower) sign refers to Fermi-Dirac (Bose-Einstein) particles. Then prove that the expression for the entropy in terms of the average occupation numbers can be written

$$S = k\sum_s [(\bar{n}_s \mp 1)\ln(1 \mp \bar{n}_s) - \bar{n}_s \ln \bar{n}_s]$$

(b) The classical limit corresponds to the case in which all $\bar{n}_s \ll 1$ (nondegenerate limit). If we set $\bar{n}_s \equiv \bar{N}f(s)$ (so that $f(s)$ represents the probability that a particle is in the state s), show that in this limit $S = -k\bar{N}\sum f(s) \ln f(s)$, plus terms involving N. Show that these terms are the same for Bose-Einstein and Fermi-Dirac particles and that they are of the same form as $C(N)$ in Problem 8 of Chapter 3. (Note: $\ln(1 + a) \simeq a$, if $a \ll 1$.)

5. The Helmholtz free energy $F = U - TS$ satisfies the thermodynamic relationship

$$dF = -S\,dT - p\,dV + \mu\,d\bar{N}$$

Therefore $dF = 0$ for any variation of an equilibrium state for which T, V, and $\bar{N}$ remain constant. Using the expression for S in Problem 4, we can write

$$F = \sum_s \epsilon(s)\bar{n}_s - kT[(\bar{n}_s \mp 1)\ln(1 \mp \bar{n}_s) - \bar{n}_s \ln \bar{n}_s]$$

The object is to show that, beginning with this expression and the above thermodynamic condition, the average occupation numbers must be given by (14.13). To show this, consider a variation of F in which T, V [hence, all $\epsilon(s)$], and N are constant. In particular let $\bar{n}_q$ and $\bar{n}_{q'}$ (for two particular states) be varied by $d\bar{n}_q$ and $d\bar{n}_{q'}$ in such a way that $d\bar{N} = 0$ (so $d\bar{n}_q = -d\bar{n}_{q'}$). Show from the above expression that $dF = 0$ only if

$$\frac{\bar{n}_q\, e^{\beta\epsilon(q)}}{1 \mp \bar{n}_q} = \frac{\bar{n}_{q'}\, e^{\beta\epsilon(q')}}{1 \mp \bar{n}_{q'}}$$

Deduce from this that $\bar{n}_q$ must be given by (14.13). [Hint: q and q' are any two states, and $dF = \mu\, d\bar{n}_q$ if only $\bar{n}_q$ is varied.]

15. THE PERFECT FERMI GAS

In this section we shall consider in more detail the properties of a system of noninteracting Fermi-Dirac particles (a perfect Fermi gas). Perhaps the most frequent application of these results is to the case of "free" electrons in solids — which we shall discuss later.

We shall first consider the properties of the distribution of particles over the various particle states $\bar{n}_q$. According to the last equation of the previous section, the **Fermi-Dirac distribution** is given by

$$\bar{n}_q = (1 + e^{-\beta[\mu - \epsilon(q)]})^{-1} \tag{15.1}$$

The parameter μ has yet to be determined in terms of the average number of particles in the system $\bar{N}$ and other variables such as the volume and the temperature. Regardless of the value of μ, notice that $\bar{n}_q$ is always less than unity in the Fermi-Dirac distribution. This, of course, must be the case since there can be at most one particle in a particle state at any instant, so the average value must be less than one.

To obtain an equation for μ we must specify the average number of particles $\bar{N}$ in the open system (just as N must be specified for the case of a closed system). Using the definition of $\bar{N}$ (see problems of the preceding section)

$$\bar{N} = \sum_q \bar{n}_q = \sum_q (1 + e^{-\beta[\mu - \epsilon(q)]})^{-1} \tag{15.2}$$

We see that this can be used to determine μ in terms of $\bar{N}$, β, and the eigenvalues $\epsilon(q)$. To actually obtain an expression for $\mu(\bar{N}, V, \beta)$ from Equation (15.2) we must use the fact that the energies $\epsilon(q)$ are very close together, so that we can replace the summation by an integral. Recall that the energy of a free particle in a cubical container with sides of length L is given by

$$\epsilon(q) = \frac{1}{2m}\left(\frac{h}{2L}\right)^2 \mathbf{q}^2 \qquad (q_x, q_y, q_z = 0, 1, 2, \ldots) \tag{15.3}$$

Moreover, if the particle has a spin† of σ, then its orientation along any axis (say the z axis) can be $\sigma_z = \sigma, \sigma - 1, \ldots, -\sigma$. That is to say, the spin can have $2\sigma + 1$ possible orientations, all of which have the same energy (15.3). Thus the letter q in (15.2) really represents four numbers, namely $(\sigma_z, q_x, q_y, q_z)$, and the summation over q becomes

$$\sum_q = \sum_{\sigma_z = -\sigma}^{\sigma} \sum_{q_x=0}^{\infty} \sum_{q_y=0}^{\infty} \sum_{q_z=0}^{\infty} \simeq \sum_{\sigma_z = -\sigma}^{\sigma} \int_0^\infty \int_0^\infty \int_0^\infty d^3q$$

Using these results, we can approximate (15.2) by

$$\bar{N} = \sum_{\sigma_z = -\sigma}^{\sigma} \int_0^\infty d^3q \left[1 + \exp\left(-\beta\mu + \left(\frac{\beta}{2m}\right)\left(\frac{h}{2L}\right)^2 \mathbf{q}^2\right)\right]^{-1} \tag{15.4}$$

Since the energy does not depend on σ_z, the summation in (15.4) simply yields $(2\sigma + 1)$ identical terms. Moreover, the quantity $(h\mathbf{q}/2L)$ is just the momentum $\mathbf{p} = m\mathbf{v}$ of the particle. Therefore, if we introduce the notation

$$g = 2\sigma + 1, \qquad \mathbf{p} = \frac{h\mathbf{q}}{2L} \tag{15.5}$$

†See footnote in previous section. The spin obeys the angular momentum rules, discussed in Section 5 of Chapter 2, with J now replaced by σ.

then (15.4) can be written

$$\bar{N} = g\left(\frac{2L}{h}\right)^3 \int_0^\infty \int_0^\infty \int_0^\infty d^3p(1 + e^{-\beta(\mu - p^2/2m)})^{-1}$$

Noting that the volume of the system is $V = L^3$, and changing the limits of integration (each integral $2\int_0^\infty$ becomes $\int_{-\infty}^\infty$ because the integrand is an even function of p_x, p_y, and p_z), we obtain

$$\bar{N} = \left(\frac{gV}{h^3}\right) \int_{-\infty}^\infty d^3p(1 + e^{-\beta(\mu - p^2/2m)})^{-1} \tag{15.6}$$

This equation determines how μ depends on $\bar{N}$, V, and T. Obviously $\mu(\bar{N}, V, T)$ is not a very simple function!

Another fairly common approach, which leads to the same result, is based on an application of Heisenberg's uncertainty principle. This principle states that a particle's position and momentum can be simultaneously determined with an accuracy Δx and Δp_x, respectively, only if $\Delta x \, \Delta p_x \geq h$. Therefore, a particle cannot be "localized" experimentally more accurately than

$$\Delta x \, \Delta y \, \Delta z \, \Delta p_x \, \Delta p_y \, \Delta p_z \geq h^3 \tag{15.7}$$

Because of this, one argues that in a region $d^3r \, d^3p$ there are $d^3r \, d^3p/h^3$ distinguishable states that a particle can occupy. If the particle has a spin σ, then there are $(2\sigma + 1) \, d^3r \, d^3p/h^3$ distinguishable states in the region $d^3r \, d^3p$, because there are $(2\sigma + 1)$ possible orientations of the spin along some axis. Using the fact that $\epsilon = p^2/2m$ and the above ideas, we can write Equation (15.2) as

$$\bar{N} = \sum_{\substack{\text{distinguishable} \\ \text{states}}} (1 + e^{-\beta(\mu - p^2/2m)})^{-1} \simeq g \int_V \int_{-\infty}^\infty \frac{d^3r \, d^3p}{h^3} (1 + e^{-\beta(\mu - p^2/2m)})^{-1}$$

where, again, $g = 2\sigma + 1$. The spatial integral yields the volume V of the system, so this expression is the same as (15.6).

In order to see the relationship between the Fermi-Dirac distribution and the classical Maxwellian distribution functions, it is useful to introduce the following form of the **Fermi-Dirac distribution.** Let

$$\boxed{\bar{n}(\mathbf{p}) = \left(\frac{gV}{h^3}\right)(1 + e^{-\beta(\mu - p^2/2m)})^{-1}} \tag{15.8}$$

Since, according to (15.6),

$$\boxed{\bar{N} = \int_{-\infty}^\infty d^3p \, \bar{n}(\mathbf{p})} \qquad [\text{determines } \mu(\bar{N}, V, T)] \tag{15.9}$$

we can see that

$$\bar{n}(\mathbf{p})\,d^3p = \begin{array}{l}\text{average (probable) number of particles in the}\\ \text{region } d^3p \text{ about } \mathbf{p}\end{array} \tag{15.10}$$

The corresponding classical distribution function is the Maxwellian distribution function (Section 2)

$$\eta(\mathbf{v})\,d^3v = \text{probable number of particles in the region } d^3v \text{ about } \mathbf{v}$$

or, if we make a trivial change of variables from the velocity to the momentum $\mathbf{p} = m\mathbf{v}$, then

$$\eta(\mathbf{p})\,d^3p = \text{probable number of particles in the region } d^3p \text{ about } \mathbf{p}$$

$$\eta(\mathbf{p}) = N(2\pi mkT)^{-3/2}e^{-\beta p^2/2m} \tag{15.11}$$

There is more apparent difference than similarity between this Maxwellian distribution and the Fermi-Dirac distribution, Equation (15.8). One difference is relatively minor and should be clarified immediately. This difference is that $\bar{n}(p)$ is, according to (15.9), normalized to $\bar{N}$, whereas $\eta(\mathbf{p})$ is normalized to N. The reason is simply that we are now using open systems rather than closed systems as in (15.11). Obviously, in comparing two such systems, we expect that they will have the same thermodynamic properties only if $N = \bar{N}$ (i.e., the open system has, on the average, the same number of particles as the closed system). This identification will be used throughout the following discussion (see problems). More significant is the obvious functional difference between $\bar{n}(\mathbf{p})$ and $\eta(\mathbf{p})$. Is there any limiting case in which these two distributions become equal to one another? To see that there is, assume that $e^{\beta\mu} \ll 1$ and note that in that case

$$\bar{n}(\mathbf{p}) = \frac{gV}{h^3}\frac{1}{1 + e^{-\beta(\mu - p^2/2m)}} = \frac{gV}{h^3}\frac{e^{\beta(\mu - p^2/2m)}}{e^{\beta(\mu - p^2/2m)} + 1} \simeq \frac{gV}{h^3}\,e^{\beta\mu}e^{-\beta p^2/2m}$$

where the last equality comes from setting $(e^{\beta(\mu - p^2/2m)} + 1)^{-1} \simeq 1$. Thus

$$\bar{n}(\mathbf{p}) \simeq \left(\frac{gV}{h^3}\right)e^{\beta\mu}e^{-\beta p^2/2m} \qquad (\text{if } 1 \gg e^{\beta\mu}) \tag{15.12}$$

Comparing (15.11) and (15.12) we find that they both depend on $\mathbf{p}$ in the same way, namely as $e^{-\beta p^2/2m}$. The coefficient of this term, however, looks different in the two cases. To see that they are actually the same, we must obtain an expression for $e^{\beta\mu}$ in terms of $\bar{N}$. If we substitute (15.12) into (15.9), we obtain

$$\bar{N} = \int_{-\infty}^{\infty} d^3p\,\bar{n}(\mathbf{p}) \simeq \left(\frac{gV}{h^3}\right)\int_{-\infty}^{\infty} d^3p\,e^{-\beta p^2/2m}e^{\beta\mu} = \left(\frac{gV}{h^3}\right)\left(\frac{2m\pi}{\beta}\right)^{3/2}e^{\beta\mu}$$

or, in other words,

$$e^{\beta\mu} \simeq \frac{h^3}{gV}\frac{\bar{N}}{(2\pi mkT)^{3/2}} \qquad (\text{if } 1 \gg e^{\beta\mu}) \tag{15.13}$$

Substituting this into (15.12), we finally obtain

$$\bar{n}(\mathbf{p}) \simeq \bar{N}(2\pi mkT)^{-3/2}\, e^{-\beta p^2/2m} \qquad (\text{if } 1 \gg e^{\beta\mu}) \tag{15.14}$$

This distribution function is now identical with the Maxwellian distribution (15.11). Therefore, we recover the classical result if $1 \gg e^{\beta\mu}$. Now what does this mean? According to (15.13) it means that

$$\boxed{1 \gg \frac{\bar{N}h^3}{gV(2\pi mkT)^{3/2}} \qquad (\text{classical limit})} \tag{15.15}$$

In Section 8 we obtained a very similar condition for a *nondegenerate perfect gas:*

$$1 \gg \frac{6}{\pi}\,\frac{Nh^3}{V(2mkT)^{3/2}} \tag{8.17}$$

Aside from unimportant factors, we see that (8.17) is identical with Equation (15.15). Therefore, we have again shown that the translational motion of a nondegenerate perfect gas can be described by a Maxwellian distribution.

Having found that the nondegenerate gas corresponds to the limit $1 \gg e^{\beta\mu}$, let us consider what would happen if $e^{\beta\mu} \gg 1$. Obviously this limit must correspond to a very degenerate gas — one that can be expected to be quite different from the classical gas. It is not so obvious that, in fact, it is possible to have $e^{\beta\mu} \gg 1$, for this requires that $\mu > 0$ and $\beta\mu \gg 1$. In the last section, however, we noted that μ can be positive in a Fermi gas, and we shall see later that in fact $\mu > 0$ as T goes to zero (so $\beta\mu \gg 1$). Therefore, we shall assume that $\beta\mu \gg 1$ as $T \to 0$ and see what happens to the Fermi-Dirac distribution in this limit. According to (15.8) the distribution function is

$$\bar{n}(\mathbf{p}) = \left(\frac{gV}{h^3}\right)(1 + e^{-\beta(\mu - p^2/2m)})^{-1} \tag{15.8}$$

and we see that for very large β (that is, $T \to 0$) the factor

$$1 + e^{-\beta(\mu - p^2/2m)} \simeq \begin{cases} 1, & \text{if } p^2/2m < \mu \\ \infty, & \text{if } p^2/2m > \mu \end{cases}$$

— in other words this factor rapidly increases from one to essentially infinity as $p^2/2m$ is increased through the value of μ. Since this value of μ is of special importance, it is useful to define the **Fermi energy** ϵ_F to be the value of $\mu(T, V, \bar{N})$ as T goes to zero:

$$\epsilon_F(V, \bar{N}) \equiv \lim_{T\to 0} \mu(T, V, \bar{N})$$

With this definition, and the above result, we conclude that the distribution function (15.8) becomes, in the limit $T \to 0$,

$$\bar{n}(\mathbf{p}) = \begin{cases} gV/h^3, & \text{if } p^2/2m < \epsilon_F \\ 0, & \text{if } p^2/2m > \epsilon_F \end{cases} \tag{15.16}$$

This distribution function is entirely different from the Maxwellian distribution $\eta(\mathbf{p})$. The function $\bar{n}(\mathbf{p})$ in (15.16) is discontinuous at momenta with the value $p^2/2m = \epsilon_F$. If $\bar{n}(\mathbf{p})$ is represented in momentum space, Figure 15.1, it has a uniform value of gV/h^3 inside a sphere of radius $\sqrt{2m\epsilon_F}$

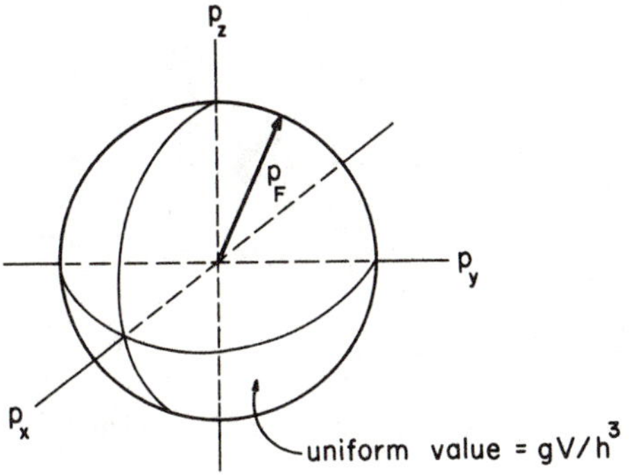

Figure 15.1. The Fermi sphere.

and is zero outside this sphere. This sphere is known as the **Fermi sphere,** and its radius is often referred to as the **Fermi momentum:**

$$p_F \equiv \sqrt{2m\epsilon_F} \tag{15.17}$$

The Fermi energy $\epsilon_F(V, \bar{N})$, which determines the radius of the Fermi sphere, is easily obtained by substituting the distribution function (15.16) into Equation (15.9):

$$\bar{N} = \int_{-\infty}^{\infty} d^3p\, \bar{n}(\mathbf{p}) \tag{15.9}$$

Since $\bar{n}(\mathbf{p})$ is constant inside the Fermi sphere, this integral equals the volume of the sphere $4\pi p_F^3/3$ times the value of $\bar{n}(\mathbf{p})$, so (15.9) becomes

$$\bar{N} = \left(\frac{4\pi}{3}\right)(2m\epsilon_F)^{3/2}\left(\frac{gV}{h^3}\right)$$

or

$$\epsilon_F(V, \bar{N}) = \frac{h^2}{2m}\left(\frac{3\bar{N}}{4\pi gV}\right)^{2/3} \tag{15.18}$$

This shows that μ is, in fact, a positive number in this degenerate limit. It should also be noted that $kT \gg \epsilon_F$ corresponds to the nondegenerate limit, Equation (15.15).

Now let us compare the Fermi-Dirac distribution in the nondegenerate limit, Equation (15.14), to the distribution in the completely degenerate limit, Equation (15.16). First, the picture analogous to Figure 15.1 in the nondegenerate (Maxwellian) limit is shown in Figure 15.2. This is essentially the same as Figure 2.1 in Section 2. In Figure 15.2 we have added

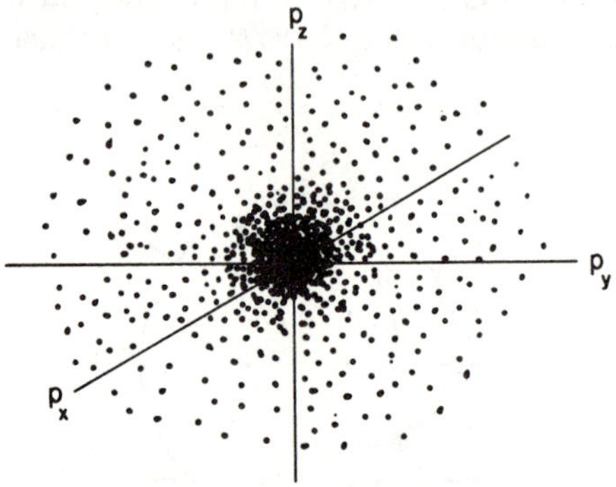

Figure 15.2. The Maxwellian distribution.

points to illustrate the fact that the distribution function decreases proportional to $e^{-\beta p^2/2m}$ (the points in Figure 15.1 would be of uniform density out to a radius p_F). Thus, as T decreases, the distribution function $\bar{n}(\mathbf{p})$ goes from the "fuzzy" Maxwellian form (Figure 15.2) to the uniform Fermi sphere (Figure 15.1). Therefore, even at $T = 0$, most of the particles have a momentum that is *not zero*! If $T = 0$, why do the particles persist in moving around? Why do they not simply settle down and rest? One would certainly think that this would represent the lowest energy state of the system. Indeed the particles would come to rest if they could, but the Fermi-Dirac particles cannot have more than one particle in each particle state — the occupation numbers for Fermi-Dirac particles can be only

$$n_s = 0, 1 \tag{15.4}$$

Therefore, when one particle occupies a state with a small value of $\mathbf{p}$, this excludes other particles from also occupying that particle state (this is referred to as the exclusion principle). Hence, as the temperature is decreased, some particles will occupy states with small values of $\mathbf{p}$, and the rest will be forced to occupy states with larger values of $\mathbf{p}$ because of the exclusion principle. The Fermi sphere is therefore the distribution with the lowest energy that, at the same time, satisfies the exclusion principle. Obviously this picture would be quite different if there were no exclusion principle, for then all the particles could crowd into the lowest energy state. This is what happens with Bose-Einstein particles — but we shall consider this case in the following section.

In order to see even more clearly the transition from the nondegenerate to the degenerate case, let us determine some of the other distribution functions related to $\bar{n}(\mathbf{p})$ and compare them with those obtained in Section 2. If we let

$$\bar{n}(p)\,dp = \begin{array}{l}\text{probable number of particles with a momentum}\\ \text{(magnitude) in the range } dp \text{ of } p\end{array} \tag{15.19}$$

then, using spherical coordinates,

$$\bar{n}(p)\,dp = \int_0^{2\pi} d\phi \int_0^{\pi} d\theta \sin\theta\, \bar{n}(\mathbf{p})p^2\,dp = 4\pi p^2 \bar{n}(\mathbf{p})\,dp$$

or

$$\boxed{\bar{n}(p) = \left(\frac{4\pi g V}{h^3}\right)\left(\frac{p^2}{1 + e^{-\beta(\mu - p^2/2m)}}\right)} \tag{15.20}$$

In the nondegenerate limit we can use (15.14) for $\bar{n}(\mathbf{p})$, whereas in the extremely degenerate limit we should use (15.16). These are illustrated (rather schematically) in Figure 15.3 by the two solid curves. At $T = 0$ the distribution function has a sharp cutoff at the Fermi momentum p_F. The fact that $\bar{n}(p)$ increases proportional to p^2 is simply due to the number of available particle states $4\pi p^2\,dp$ in the region dp about p. The dotted curve represents the distribution in a degenerate case when $T \neq 0$. Note that the particles near the Fermi surface (that is, $p \simeq p_F$) are the ones first affected by the rise in temperature. At high temperatures one recovers the non-

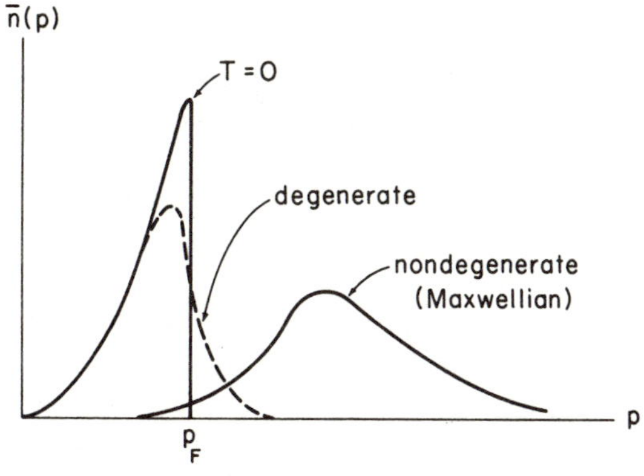

Figure 15.3

degenerate (Maxwellian) distribution (which is actually much further out on the p axis in Figure 15.3).

The probable number of particles with the x components of the momentum in the range dp_x is obtained from

$$\bar{n}(p_x)dp_x = \left[\int_{-\infty}^{\infty} dp_y \int_{-\infty}^{\infty} dp_z\, \bar{n}(\mathbf{p})\right] dp_x \tag{15.21}$$

We can evaluate this for the general distribution function $\bar{n}(\mathbf{p})$, (15.8), by first introducing cylindrical coordinates

$$p_y = \rho \cos\theta,\ p_z = \rho \sin\theta,\ dp_y\, dp_z = \rho\, d\theta\, d\rho$$

Substituting these into (15.8), we obtain

$$\begin{aligned}\bar{n}(p_x) &= \int_0^{2\pi} d\theta \int_0^{\infty} \left(\frac{gV}{h^3}\right)(1 + e^{-\beta(\mu - p_x^2/2m - \rho^2/2m)})^{-1}\rho\, d\rho \\ &= 2\pi\left(\frac{gV}{h^3}\right)\int_0^{\infty} \frac{(-m/\beta)\, d(e^{\beta(\mu - p_x^2/2m - \rho^2/2m)})}{e^{\beta(\mu - p_x^2/2m - \rho^2/2m)} + 1} \\ &= -\left(\frac{2\pi mgV}{\beta h^3}\right)\int_0^{\infty} d[\ln(e^{\beta(\mu - p_x^2/2m - \rho^2/2m)} + 1)]\end{aligned}$$

or finally

$$\boxed{\bar{n}(p_x) = \left(\frac{2\pi mgV}{\beta h^3}\right)\ln(e^{\beta(\mu - p_x^2/2m)} + 1)} \tag{15.22}$$

This is not a particularly simple distribution function, but in the limit $e^{\beta\mu} \ll 1$ it reduces to the usual Maxwellian result (see problems), and in the limit $e^{\beta\mu} \gg 1$ we can set

$$e^{\beta(\mu - p_x^2/2m)} + 1 \simeq e^{\beta(\mu - p_x^2/2m)}$$

Therefore, in the low-temperature limit

$$\bar{n}(p_x) = \left(\frac{2\pi mgV}{\beta h^3}\right)\beta\left(\mu - \frac{p_x^2}{2m}\right)$$

and since $\mu(T = 0) = \epsilon_F$ and p_x must be less than p_F, we have

$$\bar{n}(p_x) = \left(\frac{2\pi mgV}{h^3}\right)\left(\epsilon_F - \frac{p_x^2}{2m}\right) \qquad (|p_x| < p_F) \tag{15.23}$$

This distribution is illustrated in Figure 15.4. It is a parabolic curve with the characteristic sharp cutoff at the Fermi momentum. The high-temperature Maxwellian distribution is also schematically indicated in Figure 15.4. By comparison, it is very broad and much smaller in magnitude (within the Fermi sphere).

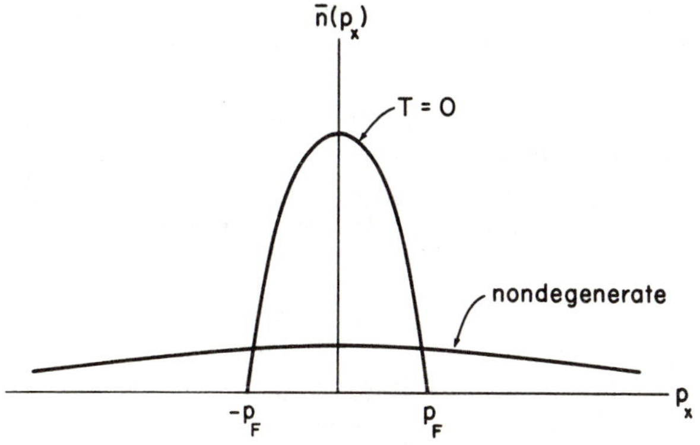

Figure 15.4

Finally, one should be warned that the distribution function $\bar{n}(\mathbf{p})$ is frequently plotted as a function of ϵ, as illustrated in Figure 15.5. This is an attempt to represent the Fermi sphere in a one-dimensional diagram (rather than as in Figure 15.1). Obviously Figure 15.5 has one advantage — it is simple to draw. One can also readily see how the Fermi sphere begins to

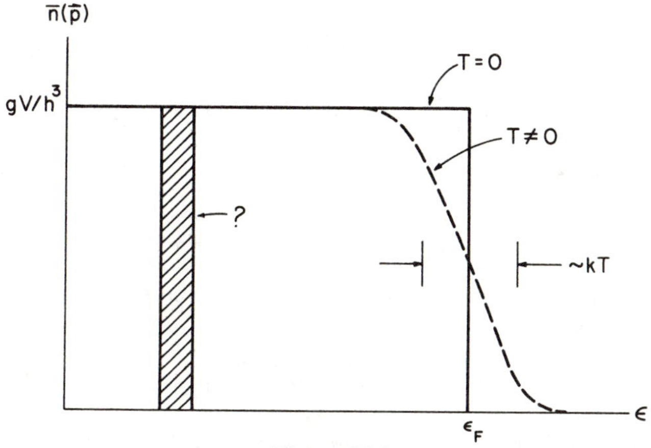

Figure 15.5

spread out over an energy range kT, when $T \neq 0°$K. However, the quantity $\bar{n}(\mathbf{p})\, d\epsilon$ (shaded area) does *not represent anything simple*. Only $\bar{n}(\mathbf{p})\, d^3p$ has some significance [Equation (15.10)]. Alternatively we can define a distribution function for the energy $\bar{n}(\epsilon)$ such that

$$\bar{n}(\epsilon)\, d\epsilon = \text{probable number of particles in the range } d\epsilon \text{ of } \epsilon \tag{15.24}$$

The corresponding figure for $\bar{n}(\epsilon)$ is not the same as Figure 15.5 (see problems). Thus Figure 15.5 is a useful diagram, provided that it is interpreted correctly.

Now let us investigate the thermodynamic properties of a perfect Fermi gas. We shall show first that there is a definite relationship between the pressure and the internal energy of a Fermi gas. To do this, we recall [Chapter 3, Equation (62)] that the pressure is related to the partition function by

$$p = kT\left(\frac{\partial \ln \mathcal{Z}}{\partial V}\right)_{\mu,T} \tag{15.25}$$

and, according to (14.10), $\ln \mathcal{Z}$ is given by

$$\begin{aligned}\ln \mathcal{Z} &= \sum_{s=1}^{\infty} \ln\left(1 + e^{\beta[\mu-\epsilon(s)]}\right)\\ &\simeq \left(\frac{gV}{h^3}\right)\int_{-\infty}^{\infty} \ln\left(1 + e^{\beta(\mu-p^2/2m)}\right) d^3p\end{aligned}$$

Here we have approximated the series by the integrals over $\mathbf{p}$ in the same way that we approximated Equation (15.2) by Equation (15.6). If we introduce spherical coordinates in the momentum variables and integrate over the angular variables, then this becomes

$$\begin{aligned}\ln \mathcal{Z} &= \left(\frac{4\pi gV}{h^3}\right)\int_0^{\infty} p^2 \ln\left(1 + e^{\beta(\mu-p^2/2m)}\right) dp\\ &= \left(\frac{4\pi gV}{3h^3}\right)\int_0^{\infty} \ln\left(1 + e^{\beta(\mu-p^2/2m)}\right) d(p^3)\end{aligned}$$

Integrating this by parts, we obtain

$$\ln \mathcal{Z} = \left(\frac{4\pi gV\beta}{3mh^3}\right)\int_0^{\infty} \frac{p^4\, e^{\beta(\mu-p^2/2m)}}{1 + e^{\beta(\mu-p^2/2m)}}\, dp \tag{15.26}$$

If we substitute (15.26) into (15.25) we obtain the *equation of state* of a perfect Fermi gas

$$p = \frac{4\pi g}{3mh^3}\int_0^{\infty} \frac{p^4\, dp}{1 + e^{-\beta(\mu-p^2/2m)}} \tag{15.27}$$

(p on the left side is, of course, the pressure, whereas the integration variable is the momentum). Unfortunately the integral in (15.27) cannot be evaluated in a closed form and must be tabulated as a function of β, μ, and m. Before we worry about this point, let us show that exactly the same integral arises in the expression for the internal energy. We obtain the internal (average) energy by evaluating (note problem 1 in the last section)

$$U = \bar{E} = \sum_s \epsilon(s)\bar{n}_s = \int_{-\infty}^{\infty} \left(\frac{p^2}{2m}\right)\bar{n}(\mathbf{p})\, d^3p \tag{15.28}$$

Using the general expression for $\bar{n}(\mathbf{p})$, Equation (15.8), we obtain

$$U = \frac{gV}{2mh^3}\int_{-\infty}^{\infty}\frac{p^2\,d^3p}{1 + e^{-\beta(\mu - p^2/2m)}} = \frac{2\pi gV}{mh^3}\int_0^{\infty}\frac{p^4\,dp}{1 + e^{-\beta(\mu - p^2/2m)}} \tag{15.29}$$

where we again have introduced spherical coordinates. Comparing (15.29) with (15.27), we see that the pressure and the internal energy are related by

$$\boxed{pV = \tfrac{2}{3}U} \tag{15.30}$$

This relationship can be easily verified in the nondegenerate limit, for then $U = \frac{3}{2}\bar{N}kT$ and $pV = \bar{N}kT$.

From these results it can be seen that all the thermodynamic properties are determined by the (single) integral in the above expressions for U, p, and $\ln Z$ together with the integral determining $\mu(\bar{N}, V, T)$, Equation (15.9). However, as we have already noted, these integrals cannot be evaluated in closed form. They may be expressed in terms of infinite series in the two cases $e^{\beta\mu} > 1$ (strong degeneracy) and $e^{\beta\mu} < 1$ (weak degeneracy). The manipulations that yield such series will be omitted here,† and we shall simply record the resulting first few terms in each case:

$$U = \tfrac{3}{2}\bar{N}kT\left[1 + \frac{\bar{N}h^3}{16Vg(\pi mkT)^{3/2}} + \cdots\right] \qquad \text{(weak degeneracy)} \tag{15.31}$$

$$U = \tfrac{3}{5}\bar{N}\epsilon_F\left[1 + \frac{5\pi^2}{12}\left(\frac{kT}{\epsilon_F}\right)^2 + \cdots\right] \qquad \text{(strong degeneracy)} \tag{15.32}$$

Equation (15.31) gives the first quantum correction to the classical value of the internal energy of a monatomic perfect gas $\frac{3}{2}\bar{N}kT$. When (15.31) is substituted into (15.30), one obtains the corresponding first correction to the equation of state $pV = \bar{N}kT$. Equation (15.32) gives the internal energy in the opposite limit of strong degeneracy. The first term

$$U = \tfrac{3}{5}\bar{N}\epsilon_F \qquad (T = 0°\text{K}) \tag{15.33}$$

which represents the internal energy when $T = 0°$K, can be obtained quite readily from our previous results (see problems). This result is, of course, entirely different from what one would expect from classical statistics (where $U = \frac{3}{2}\bar{N}kT = 0$, for $T = 0°$K).

To illustrate this difference, assume that the free (conduction) electrons in a metal can be treated as a perfect Fermi gas. If we assume that in silver

†See, for example, K. Huang, *Statistical Mechanics* (New York: John Wiley & Sons, Inc., 1963), Sec. 11.1, or L. D. Landau and E. M. Lifshitz, *Statistical Physics* (Reading, Mass.: Addison-Wesley Publishing Co., Inc., 1958), Sec. 57.

there is one free electron per atom, then $\bar{N}/V = 5.87 \times 10^{22}\ \text{cm}^{-3}$. Since $g = 2$ for electrons ($\sigma = \frac{1}{2}$), we find that the Fermi energy equals

$$\epsilon_F = \left(\frac{h^2}{2m_e}\right)\left(\frac{3\bar{N}}{8\pi V}\right)^{2/3} = \frac{(6.62 \times 10^{-27})^2(7 \times 10^{21})^{2/3}}{1.82 \times 10^{-27}} = 8.8 \times 10^{-12}\ \text{erg}$$

Therefore, according to (15.33), the average kinetic energy of an electron at $T = 0°\text{K}$ is $\frac{3}{5}\epsilon_F = 5.28 \times 10^{-12}$ erg. This corresponds to a velocity of approximately 2.5×10^8 cm/sec — which is quite different from the classical picture of the electron at rest when $T = 0°\text{K}$! Perhaps even more impressive is the pressure generated by this electron gas. Substituting (15.33) into (15.30) yields

$$p = \tfrac{2}{5}\left(\frac{\bar{N}}{V}\right)\epsilon_F \qquad (T = 0°\text{K}) \tag{15.34}$$

Substituting in the above values for silver, we find that the electron-gas pressure is

$$p = 2.07 \times 10^{10}\ \text{dynes/cm}^2 \simeq 20{,}000\ \text{atmospheres!}$$

This very substantial pressure is, of course, not observed outside of the metal (why?).

Of more thermodynamic significance is the second term in (15.32), which gives the low-temperature heat capacity of a Fermi gas:

$$\boxed{C_V = \left(\frac{\partial U}{\partial T}\right)_V = \left(\frac{\pi^2 kT}{2\epsilon_F}\right)\bar{N}k \qquad (\epsilon_F \gg kT)} \tag{15.35}$$

This shows that the heat capacity of the free electrons in a solid goes to zero proportional to T. The reason is that only the electrons near the Fermi surface are affected when heat is added to the system (e.g., see Figure 15.5). The number of electrons that alter their energy is proportional to kT/ϵ_F and each electron should contribute approximately $3k/2$ to the heat capacity, so C_V should be proportional to k^2T/ϵ_F.

This result explains one of the baffling questions left unanswered by classical statistical theories. According to the classical equipartition law the internal energy of these conduction electrons should be $\frac{3}{2}\bar{N}kT$, so they should contribute a heat capacity $C_V = \frac{3}{2}\bar{N}k$ to the solid. On the other hand, it was known that the heat capacity of solids at high temperatures is approximately given by the Dulong-Petit law, $C_V \simeq 3N_ak$ (N_a = number of atoms). Since this heat capacity is due to the motion of the lattice atoms (Section 10), it follows that the heat capacity of the free electrons must be much less than the classical value ($\frac{3}{2}\bar{N}k$). The result (15.35) shows that the electronic heat capacity should indeed be much less than the classical value for all temperatures $T \ll \epsilon_F/k$ (for example, $\epsilon_F/k = 64{,}000°\text{K}$ for silver, so this inequality is always satisfied).

Even though the electronic heat capacity (15.35) is very small at low temperatures, it may equal or exceed the heat capacity due to the lattice vibrations of the solid. It will be recalled (see Section 10) that the lattice heat capacity at low temperatures is proportional to T^3 (the Debye T^3 law). If we add to this the electronic heat capacity, which is proportional to T, then the total heat capacity of the solid at low temperatures should be of the form

$$\begin{aligned} C_V &= \gamma T + \alpha T^3 \\ \frac{C_V}{T} &= \gamma + \alpha T^2 \end{aligned} \tag{15.36}$$

which is sometimes referred to as the Debye-Sommerfeld equation. According to this, if C_V/T is plotted against T^2, one should obtain a straight line whose intercept at $T = 0$ yields the value of γ. Some experimental values that illustrate this fact are given in the problems.

PROBLEMS

1. Consider the electron gas in tungsten (atomic weight = 183.8). Assume that the density of tungsten is 19.3 gm/cm³ and that there are two free electrons per atom. Determine the free-electron density in the metal and the Fermi energy ϵ_F. Is this electron gas degenerate for all temperatures below the melting point (approximately 3700°K)?
2. The electron density in a discharge tube is 10^{12} per cm³. Determine the value of the characteristic temperature ϵ_F/k and explain its significance.
3. Use the fact that $\ln(1 + x) \simeq x$ (if $1 \gg x$) to obtain the Maxwellian distribution function $\eta(p_x)$ from (15.22) in the case $1 \gg e^{\beta\mu}$.
4. Plot $p_F \bar{n}(p)/\bar{N}$ vs. p/p_F for the case of the free electrons in silver for $T = 0$°K and $T = 1000$°K ($\epsilon_F = 8.8 \times 10^{-12}$ erg and $\mu \simeq \epsilon_F\left[1 - \frac{\pi^2}{12}\left(\frac{kT}{\epsilon_F}\right)^2\right] \simeq \epsilon_F$ for both cases).
5. Using the same reasoning as in Section 3, determine the probable number of Fermi particles that strike a unit area of a wall per second when $T = 0$°K. Take $g = 2$ (e.g., electrons) and express your answer in terms of $\bar{N}$, V, h, and m.
6. The conduction electrons can escape from a metal (thermionic emission) provided that their kinetic energy *normal* to the surface is greater than some minimum value ϵ_m. The energy $\epsilon_m - \mu \equiv \phi$ is known as the work function, and usually $\phi \gg kT$. Taking the surface normal to be in the x direction, obtain an expression for the emitted current per unit area in terms of $\bar{n}(p_x)$. Using the fact that $\phi \gg kT$, show that the current per unit area is of the form $I = AT^2e^{-\phi/kT}$ (the Richardson-Dushman equation). Determine the value of A. Determine the experimental value of A from the observed value of 3000 amp/m² for tungsten ($\phi = 4.52$ ev) at $T = 2500$°K. Is this elementary theory of theriomnic emission very good?

7. (a) Using the definition (15.24) for the energy distribution function $\bar{n}(\epsilon)$ and the fact that $\epsilon = p^2/2m$, obtain $\bar{n}(\epsilon)$ from $\bar{n}(p)$. Plot $\bar{n}(\epsilon)$ for $T = 0°K$, and note how it differs from Figure 15.5. Indicate $\epsilon = \bar{\epsilon}$ (the average energy of a particle) in this figure.
(b) Determine $\sigma(\epsilon)/\bar{\epsilon}$ of a single Fermi particle at $T = 0°K$, using the probability distribution function $\bar{N}^{-1}\bar{n}(\epsilon) \equiv F(\epsilon)$.

8. In Chapter 3, Section 6, it was shown that $\sigma(E)/U = (3N/2)^{-1/2}$ for a classical perfect gas. Note that this fractional standard deviation is independent of T. Determine $\sigma(E)/U$ for a degenerate Fermi gas and note its dependence on T. [Hint: Use C_V, Equation (15.35).]

9. In order for the average number of particles in an open system $\bar{N}$ to have any thermodynamic significance (so that we can identify it with a closed system, for which $N = \bar{N}$), the standard deviation $\sigma(N)$ must be much less than $\bar{N}$.
(a) Using an analysis very similar to the one in Section 6 of Chapter 3, show that

$$\sigma^2(N) \equiv \text{var}(N) = \beta^{-1}\left(\frac{\partial \bar{N}}{\partial \mu}\right)_{V,T}$$

(b) Using Equations (15.13) and (15.18), determine $\sigma(N)/\bar{N}$ in both the nondegenerate and extreme degenerate limits. What *single* condition is sufficient to insure that this ratio is small in both limits?

10. (a) Show that the internal energy at $T = 0°K$, Equation (15.33), can be obtained directly from Equations (15.28) and (15.16).
(b) Use the geometry and the method of Section 3 to obtain an expression for the momentum transfer to a wall, per unit time and area, in terms of $\bar{n}(p_x)$. Evaluate this at $T = 0°K$, using (15.23), and show that it yields (15.34).

11. We have frequently noted that a constant may always be added to the internal energy without changing the physical situation. On the other hand, of course, we cannot change the volume or pressure without changing the physical situation. How is it possible, therefore, to have a relationship such as Equation (15.30)? What has been done in the process of obtaining this equation?

12. Copper has an atomic weight of 63.5 and a density of 9 gm/cm³. Experimental values of the specific heat of copper at low temperatures are:

T (°K)	1	2	3	4	6	8
$C_p/100$ (ergs/gm-deg)	1.2	2.8	5.3	9.1	23	47

At these temperatures C_p and C_V are essentially equal.
(a) Plot $C_V/100T$ vs. T^2 and determine the constants γ and α in the Debye-Sommerfeld equation (15.36).
(b) Assume that copper has one free electron per atom. Use the value of γ from part (a), together with Equation (15.35), to determine the Fermi energy of copper.
(c) Compute the Fermi energy of copper directly from (15.18). In order to make this agree with part (b), what "effective" electron mass m_e^* must be used in Equation (15.18)? (The fact that the electron in a solid has an effective mass different from m_e is a result of the complex internal fields within the solid.)

(d) Use the value of α in part (b), together with Debye's T^3 law, Equation (10.16), to obtain the Debye temperature of copper. Compare this with the value quoted in Table 10.1 (which was obtained from higher-temperature values of C_V).

13. (a) Obtain an approximate expression for ln $\mathcal{Z}$, Equation (15.26), in the nondegenerate limit $1 \gg e^{\beta\mu}$. Using the result of problem 2, Section 14, obtain the quantum expression for the particle partition function $\mathfrak{z}$ and compare this with the classical expression

$$\mathfrak{z}_{\text{class}} = \int_V d^3r \int_{-\infty}^{\infty} d^3p\, e^{-\beta p^2/2m} = V\left(\frac{2\pi kT}{m}\right)^{3/2}$$

(b) In Chapter 3 it was noted that in order for the classical partition function to be dimensionless, it must be written in the form

$$Z = D_N \int_V d^3r_1 \cdots \int_{-\infty}^{\infty} d^3p_N\, e^{-\beta E}$$

where D_N has the dimension $(\text{length})^{-3N}$ $(\text{momentum})^{-3N}$. In a perfect gas system this becomes $Z = D_N\, \mathfrak{z}_{\text{class}}^N$. We set $D_N = 1$ because its value is immaterial to any measurable property of the system. Using the nondegenerate approximation $Z = \mathfrak{z}^N/N!$ [Equation (8.12)], together with $\mathfrak{z}$ from part (a), determine the value of D_N required by quantum mechanics. When this value of D_N is used in the classical expression, it is often referred to as "corrected" Boltzmann statistics.

(c) In terms of indistinguishability and the Heisenberg uncertainty principle, explain why quantum mechanics introduces this value for D_N.

16. THE PERFECT BOSE GAS

Because more than one Bose-Einstein particle can be in a particle state, the physical properties of a Bose-Einstein gas are quite different from those of a Fermi-Dirac gas — at least at low temperatures. At low temperatures the gas particles will be preferentially distributed over the lowest energy states. As we have seen, the particles of a Fermi-Dirac gas cannot all collect in the lowest energy state as $T \longrightarrow 0$ but are forced to form the "Fermi sphere" distribution because of the exclusion principle. Since this principle does not apply for Bose-Einstein particles, it is clear that they will tend to collect in the lowest energy state ($\epsilon = 0$) as $T \longrightarrow 0$. Since this state is one of essentially zero momentum, it is also clear that the pressure of the gas will tend to zero (as compared with the large Fermi-Dirac pressure at $T = 0°$K). What is not so clear is the manner in which the particles will begin to occupy the state $\epsilon = 0$ as the temperature is decreased. For example, it would seem reasonable to expect that the probable number of particles in the lowest state would simply increase in a smooth fashion as the temperature decreased. For large systems, at least, this expectation turns out to be incorrect. Instead, what happens is that the occupation of the state $\epsilon = 0$ increases very abruptly at a certain critical temperature, leading to an abrupt change in the behavior of all thermodynamic properties. It is this abrupt behavioral change that is really unexpected — and hence the most interesting feature of the perfect Bose-Einstein gas.

To see how this comes about, consider the average occupation number in the particle state q [with energy $\epsilon(q)$]. According to Equation (14.13)

$$\bar{n}_q = (e^{-\beta[\mu-\epsilon(q)]} \pm 1)^{-1} \qquad \begin{cases} +\text{: Fermi-Dirac} \\ -\text{: Bose-Einstein} \end{cases} \tag{16.1}$$

The only formal difference between these two distributions is the term ± 1. In the nondegenerate limit $e^{-\beta\mu} \gg 1$ this term is unimportant, and both distributions reduce to the classical Maxwellian distribution. On the other hand, if $e^{-\beta\mu} \gtrsim 1$ (i.e., $\beta\mu$ is a small negative number), then the two distributions in (16.1) are very different — particularly for the state $\epsilon = 0$.† If $\epsilon = 0$ (say for the state $q = 0$), then

$$\bar{n}_0 = (e^{-\beta\mu} \pm 1)^{-1}$$

and if $e^{-\beta\mu-} \simeq 1$, then this equals $\frac{1}{2}$ for Fermi-Dirac particles, but it is nearly infinite for Bose-Einstein particles. Because of this, and the fact that the $\bar{n}_q$ are normalized to $\bar{N}$ (a finite number) by

$$\bar{N} = \sum_q \bar{n}_q \tag{16.2}$$

it follows that, unlike Fermi-Dirac particles, μ must always be negative for Bose-Einstein particles. This point was also noted in Section 14. From these considerations it is evident that the occupation number for the state $\epsilon = 0$ will have to be treated carefully in the degenerate limit $\beta\mu \lesssim 0$.

First let us see what happens if we do not treat this state carefully. If we proceed as in the case of Fermi-Dirac particles, we simply replace the summation in (16.2) by an integral and obtain an equation analogous to (15.6):

$$\bar{N} = \left(\frac{gV}{h^3}\right) \int_{-\infty}^{\infty} (e^{-\beta(\mu-p^2/2m)} - 1)^{-1}\, d^3p \tag{16.3}$$

This equation should presumably be used to determine $\mu(\bar{N}, V, T)$. For a fixed value of $\bar{N}$ and V we know that μ is large and negative when T is large (the nondegenerate limit). As T decreases, Equation (16.3) yields a value of μ that is a smaller and smaller negative number (shown schematically in

†That is to say, the particle state of lowest energy — which we can take to be zero.

Figure 16.1). For some temperature $T = T_c$, Equation (16.3) will yield the value $\mu = 0$. This happens when

$$\bar{N} = \left(\frac{gV}{h^3}\right)\int_{-\infty}^{\infty} (e^{p^2/2mkT_c} - 1)^{-1}\, d^3p$$

$$= \left(\frac{4\pi gV}{h^3}\right)\int_{0}^{\infty} (e^{p^2/2mkT_c} - 1)p^2\, dp$$

$$= \left(\frac{4\pi gV}{h^3}\right)(2mkT_c)^{3/2}\int_{0}^{\infty} \frac{x^2\, dx}{e^{x^2} - 1}$$

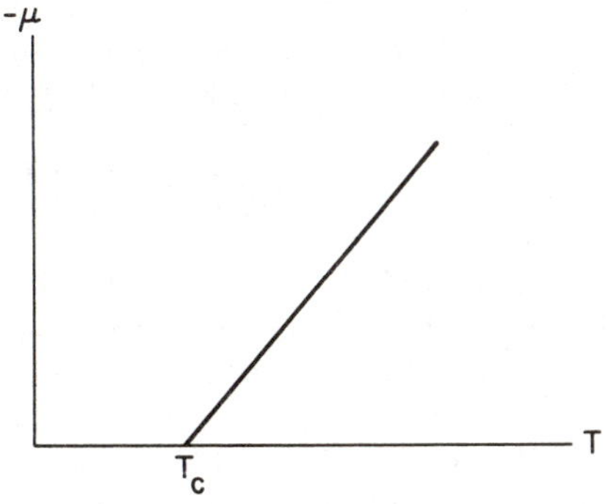

Figure 16.1

The last integral has the value $2.612\pi^{1/2}/4$ (see problems). Therefore, (16.3) predicts that $\mu = 0$ when $T = T_c$, which defines a "critical temperature"

$$T_c = \frac{h^2}{2\pi mk}\left(\frac{N}{2.612gV}\right)^{2/3} = \frac{0.0837h^2}{mk}\left(\frac{\bar{N}}{gV}\right)^{2/3} \tag{16.4}$$

Now there is obviously something wrong with the conclusion that $\mu = 0$ when $T = T_c$—for we know that μ cannot be zero in the original series

$$\bar{N} = \sum_{q} (e^{-\beta[\mu - \epsilon(q)]} - 1)^{-1} \tag{16.2}$$

because the term $\epsilon = 0$ would be infinite. We conclude, therefore, that the integral in (16.3) does not accurately represent the series in (16.2) when $\mu \simeq 0$. What we must do is first remove the singular term ($\epsilon = 0$) from the series (16.2); then we can represent the remaining terms by the integral in (16.3). Doing this, we obtain

$$\bar{N} = (e^{-\beta\mu} - 1)^{-1} + \left(\frac{gV}{h^3}\right)\int_{-\infty}^{\infty} (e^{-\beta(\mu - p^2/2m)} - 1)^{-1}\, d^3p \tag{16.5}$$

The first term on the right side of (16.5) represents the average occupation number in the lowest energy state $\bar{n}_0$ and the integral term equals the number of particles in all of the other states.

It is quite apparent that Equation (16.5), which determines $\mu(\bar{N}, V, T)$, is not a very simple expression. Indeed, the only way to solve (16.5) in general is by numerical methods. However, if we consider the important limiting case

$$\bar{N} \longrightarrow \infty, V \rightarrow \infty, \qquad \frac{N}{V} = \text{constant} \tag{16.6}$$

which corresponds to a very large macroscopic system with a fixed number density, then (16.5) yields fairly simple and very interesting results. To make this clearer, divide (16.5) by N and write it in the form

$$1 = \left(\frac{\bar{n}_0}{\bar{N}}\right) + \left(\frac{gV}{\bar{N}h^3}\right) \int_{-\infty}^{\infty} (e^{-\beta(\mu - p^2/2m)} - 1)^{-1}\, d^3p \tag{16.7}$$

where $\bar{n}_0 = (e^{-\beta\mu} - 1)^{-1}$. Now if $T > T_c$, then the solution of (16.7) is the same as (16.3), when we use the limit (16.6). This is so because the term $\bar{n}_0/\bar{N}$ in (16.7) goes to zero when $\bar{N} \to \infty$, provided that $\bar{n}_0$ does not also become infinite. Assuming that $\bar{n}_0$ remains finite, Equation (16.7) reduces to (16.3) in the limit (16.6). The resulting value of μ (Figure 16.1) shows that $\bar{n}_0$ is indeed finite, as assumed. What all this means is that, if $T > T_c$, the number of particles in the lowest energy state $\bar{n}_0$ is a vanishingly small fraction of the total number of particles $\bar{N}$ [in the limit (16.6)]. If, on the other hand, $T < T_c$, then $\bar{n}_0/\bar{N}$ cannot vanish in the limit (16.6). If it did vanish, we would again obtain (16.3) — which would have no solution. To determine the value of $\bar{n}_0/\bar{N}$ when $T < T_c$, we note that μ must be very small, and hence we can set it equal to zero in the integral term in (16.7) (but not in the term $\bar{n}_0$, of course). This yields

$$\begin{aligned} 1 &= \left(\frac{\bar{n}_0}{\bar{N}}\right) + \left(\frac{gV}{\bar{N}h^3}\right) \int_{-\infty}^{\infty} (e^{\beta p^2/2m} - 1)^{-1}\, d^3p \\ &= \left(\frac{\bar{n}_0}{\bar{N}}\right) + \left(\frac{4\pi gV}{\bar{N}h^3}\right) (2mkT)^{3/2} \int_0^{\infty} \frac{x^2\, dx}{e^{x^2} - 1} \end{aligned}$$

Referring back to (16.4) and the preceding equation, we see that this equals

$$1 = \left(\frac{\bar{n}_0}{\bar{N}}\right) + \left(\frac{T}{T_c}\right)^{3/2}$$

so

$$\boxed{\bar{n}_0 = \bar{N}\left[1 - \left(\frac{T}{T_c}\right)^{3/2}\right] \qquad (T \le T_c)} \tag{16.8}$$

This result shows that at $T = 0°\text{K}$ all the particles are in the lowest energy state ($p^2/2m = 0$, or $p = 0$). As T increases, the number in this state decreases (see Figure 16.2), and the remainder are "smeared out" over all the other particle states. These particles can be represented by a (continuous) distribution function such as

$$\bar{n}(p) = \frac{\left(\frac{4\pi gV}{h^3}\right) p^2}{(e^{-\beta(\mu - p^2/2m)} - 1)^{-1}} \tag{16.9}$$

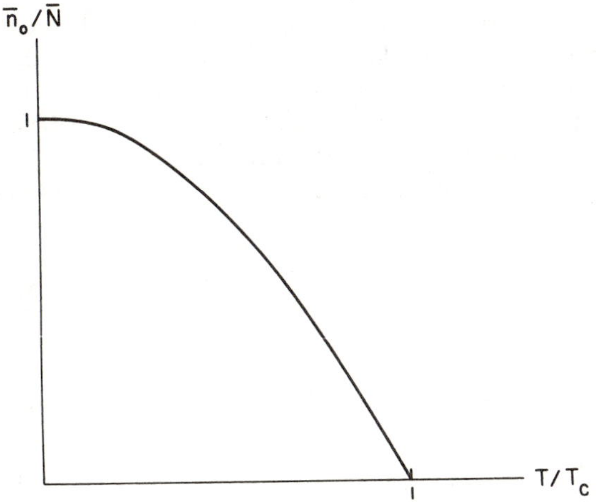

Figure 16.2

Furthermore, we can set $\mu = 0$ in (16.9) when $T < T_c$. Notice that $\bar{n}(p = 0) = 0$ is *not* the same as $\bar{n}_0$. The latter represents the occupation number of *one state* (the lowest energy state), whereas $\bar{n}(p)$ is an ordinary distribution function such that $\bar{n}(p)\,dp$ = average number of particles in the range dp about p. The distribution of particles over the momentum states (16.9) is indicated in Figure 16.3, where we have included the singular

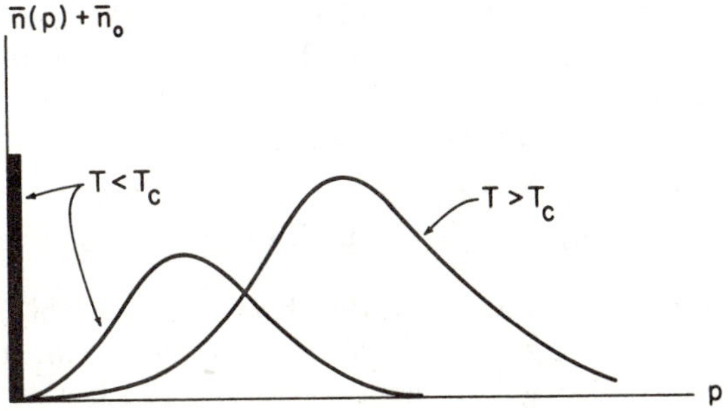

Figure 16.3

part $\bar{n}_0$. For temperatures below the critical temperature T_c there is a finite fraction of the particles in the single lowest-momentum state. This phenomenon is often referred to as **Bose-Einstein condensation** — the particles "condense" into the state $p = 0$.

This condensation of the particles into the state $p = 0$ produces abrupt changes in the thermodynamics properties when T decreases through the critical value T_c. The particles that are in the state $p = 0$ do not contribute to the pressure or internal energy of the system (they have zero momentum), so the pressure and U rapidly decrease as $T \to 0$ (see problems). One of the most interesting results of this condensation is the abrupt change in the heat capacity at $T = T_c$. This is indicated in Figure 16.4. At temperatures well

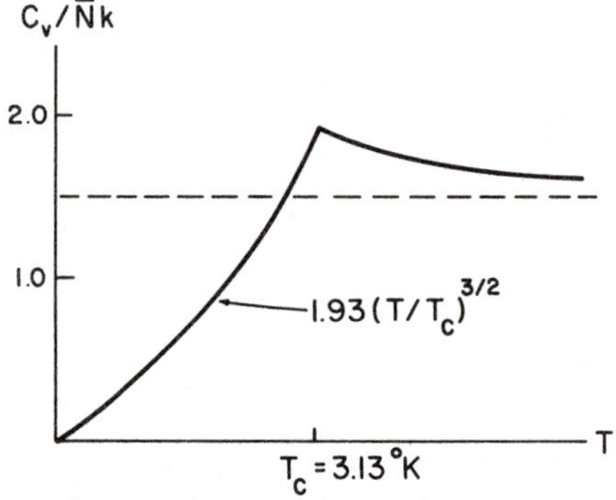

Figure 16.4

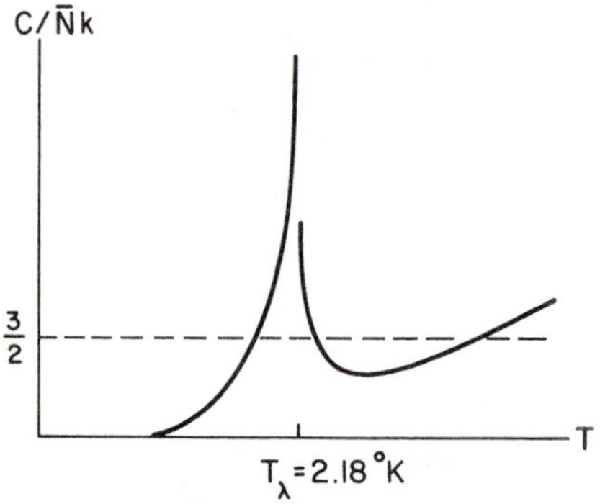

Figure 16.5

above T_c the heat capacity has the classical value of $\frac{3}{2}\bar{N}k$, and it increases as $T \to T_c$. At T_c it abruptly changes its slope $(\partial C_V/\partial T)$ because of the loss of particles to the lowest energy state. The critical temperature (16.4), at which this should occur, depends inversely on the mass of the Bose-Einstein particle. Thus the temperature will be the highest (and most easily observed) if the particle mass is small. The Bose-Einstein particle with the smallest mass, which does not solidify at low temperatures, is He^4. According to (16.4), the critical temperature for He^4 should be

$$T_c = \frac{0.0837(6.62 \times 10^{-27})^2}{4 \times 1.66 \times 10^{-24} \times 1.38 \times 10^{-16}} \left(\frac{0.1466}{4 \times 1.66 \times 10^{-24}}\right)^{2/3} = 3.13°\text{K}$$

where we have set $g = 1$ and used the observed density of He^4 in this temperature range (0.1466 gm/cm³). Of course, at these temperatures He^4 is a liquid and not a perfect gas, so that the results of this section are not strictly applicable (see problems). Nonetheless, the observed heat capacity of He^4, shown in Figure 16.5, does bear a vague resemblance to Figure 16.4. The heat capacity of liquid He^4 obviously exhibits a more spectacular rise at the transition point than does the perfect Bose-Einstein gas. However, the critical temperature computed above is fairly close to the "λ-point" temperature (2.18°K) of He^4 (the name "λ point" comes from the form of Figure 16.5). Another reason for the belief that the λ point in liquid He^4 is related in some fashion to a Bose-Einstein condensation is that no transition of this type has been observed in He^3 (consisting of Fermi-Dirac particles). Since the interatomic forces within liquid He^4 and He^3 should be quite similar, it appears that the only significant difference between these systems is that one consists of Bose-Einstein particles and the other contains Fermi-Dirac particles. This is a strong indication that the λ point is related to a Bose-Einstein condensation. At present, however, no theory has been devised for adequately treating the interactions between the He^4 atoms in the liquid. Presumably, this should account for the difference between Figures 16.4 and 16.5.

PROBLEMS

1. What is the difference in the areas under the two continuous curves in Figure 16.3? Explain your answer.
2. Radiation can be thought of as a perfect Bose gas of photons (with $\epsilon = cp$ being the relationship between energy and momentum). Why doesn't the photon gas exhibit Bose condensation? In other words, what condition is imposed in the present section (but not for radiation) that requires the condensation?
3. Consider the integral ($\mu \leq 0$)

$$\int_{-\infty}^{\infty} \frac{d^3p}{e^{-\beta(\mu-p^2/2m)} - 1} = 4\pi \int_0^{\infty} \frac{p^2\, dp}{e^{-\beta(\mu-p^2/2m)} - 1}$$

Multiply the numerator and denominator by $e^{\beta(\mu-p^2/2m)}$ and use the fact that

$(1 - x)^{-1} = \sum_{n=0}^{\infty} x^n$, $(x < 1)$, to prove that this integral equals

$$(2\pi mkT)^{3/2} \sum_{n=0}^{\infty} \frac{e^{(n+1)\beta\mu}}{(n+1)^{3/2}}$$

[the series $\sum_{n=0}^{\infty} (n+1)^{-3/2} = 2.612\ldots$ quoted in the text].

4. The thermodynamic properties of a Bose-Einstein perfect gas below the critical temperature can be determined fairly easily. When $T < T_c$, the distribution function (16.9) may be used with $\mu = 0$. Note again that the "condensed" particles do not contribute to the internal energy or the pressure of the system.
(a) Using a procedure analogous to the one in problem 3, obtain an expression for the internal energy U of the Bose-Einstein gas when $T < T_c$ [note $\sum_{n=0}^{\infty} (n+1)^{-5/2} = 1.342\ldots$]. From this compute C_V and compare it with the result shown in Figure 16.4.
(b) The relationship between the pressure and the internal energy (15.30) is the same for either a Bose-Einstein or a Fermi-Dirac gas. Using the results of (a), obtain the equation of state for $T < T_c$, and note how p depends on V for constant T.
5. Do problem 5 in Section 10.

Appendices

A. GAUSSIAN INTEGRALS

Integrals involving the Gaussian function e^{-x^2} often occur in statistical mechanics. We shall now show how these integrals may be evaluated. Consider first the integral

$$I = \int_{-\infty}^{\infty} e^{-x^2}\,dx$$

To evaluate this, we use a little trick. If we consider I^2, and transform to cylindrical coordinates, then we obtain

$$I^2 = \int_{-\infty}^{\infty} e^{-x^2}dx \int_{-\infty}^{\infty} e^{-y^2}dy = \int_0^{\infty} r\,dr \int_0^{2\pi} d\theta\, e^{-r^2} = 2\pi(-\tfrac{1}{2}e^{-r^2})_0^{\infty} = \pi$$

Thus we find that $I = \sqrt{\pi}$. By the substitution $\beta y^2 = x^2$ we also find that

$$\int_{-\infty}^{\infty} e^{-\beta y^2}\,dy = \sqrt{\frac{\pi}{\beta}}$$

and since the integrand is an even function of y, the integral of $e^{-\beta y^2}$ from $-\infty$ to 0 must equal the integral from 0 to $+\infty$, so

$$G_0 = \int_0^\infty e^{-\beta y^2} dy = \frac{1}{2}\sqrt{\frac{\pi}{\beta}}$$

To evaluate integrals of the form

$$G_{2n} = \int_0^\infty y^{2n} e^{-\beta y^2}\, dy \qquad (n = 1, 2, \ldots)$$

we note that, for example,

$$\frac{\partial G_0}{\partial \beta} = \int_0^\infty -y^2 e^{-\beta y^2}\, dy = -G_2$$

But

$$\frac{\partial G_0}{\partial \beta} = \frac{\partial(\frac{1}{2}\sqrt{\pi/\beta})}{\partial \beta} = \frac{-\frac{1}{4}\sqrt{\pi}}{(\beta)^{3/2}}$$

hence

$$G_2 = \frac{\frac{1}{4}\sqrt{\pi}}{(\beta)^{3/2}}$$

Similarly

$$G_{2n} = (-1)^n \frac{\partial^n}{\partial \beta^n}\left(\frac{1}{2}\sqrt{\frac{\pi}{\beta}}\right) = \frac{1\cdot 3\cdot 5\cdots(2n-1)}{2^{n+1}}\sqrt{\frac{\pi}{\beta^{2n+1}}} \qquad \text{(A.1)}$$

On the other hand, integrals of the form

$$G_{2n+1} = \int_0^\infty y^{2n+1} e^{-\beta y^2}\, dy$$

can be evaluated by noting that

$$G_1 = \int_0^\infty y e^{-\beta y^2}\, dy = -\frac{1}{2\beta}\int_0^\infty d(e^{-\beta y^2}) = \frac{1}{2\beta}$$

As before, we can generate all G_{2n+1} from G by differentiating. Thus

$$G_{2n+1} = (-1)^n \frac{\partial^n}{\partial \beta^n}\left(\frac{1}{2\beta}\right) = \frac{n!}{2\beta^{n+1}} \qquad \text{(A.2)}$$

Finally, we note that the "displaced" Gaussian $e^{-\beta(x-x_0)^2}$ can be evaluated by a change in variables. Thus

$$\int_{-\infty}^\infty y^n e^{-\beta(y-y_0)^2}\, dy = \int_{-\infty}^\infty (x + y_0)^n e^{-\beta x^2}\, dx$$

Once the integrand $(x + y_0)^n$ is expanded, we can integrate the individual terms using Equations (A.1) and (A.2).

B. THE ERROR FUNCTION

$$\text{erf}(x) = (2/\sqrt{\pi}) \int_0^x e^{-u^2}\, du$$

x	erf (x)	x	erf (x)
0.10	0.11246	1.60	0.97635
0.20	0.22270	1.70	0.98379
0.30	0.32863	1.80	0.98909
0.40	0.42839	1.90	0.99279
0.50	0.52050	2.00	0.99532
0.60	0.60386	2.10	0.99702
0.70	0.67780	2.20	0.99814
0.80	0.74210	2.30	0.99886
0.90	0.79691	2.40	0.99931
1.00	0.84270	2.50	0.99959
1.10	0.88021	2.60	0.99976
1.20	0.91031	2.70	0.99987
1.30	0.93401	2.80	0.99992
1.40	0.95229	2.90	0.99996
1.50	0.96611	3.00	0.99998

For all values of x

$$\text{erf}(x) = \frac{2}{\sqrt{\pi}}\left(x - \frac{x^3}{3\cdot 1!} + \frac{x^5}{5\cdot 2!} - \frac{x^7}{7\cdot 3!} + \cdots\right)$$

For large positive x one can use the series

$$\text{erf}(x) \backsim 1 - \frac{e^{-x^2}}{x\sqrt{\pi}}\left(1 - \frac{1}{2x^2} + \frac{1\cdot 3}{(2x^2)^2} - \cdots\right)$$

C. AN ALTERNATIVE DERIVATION OF $\beta = 1/kT$

In Chapter 2 the parameter β was identified by a method that made use of the partition function (and internal energy) of a perfect gas. Since β is independent of the composition of the system, this identification, once made, holds in general for any system. In the present appendix we shall consider another method of identifying β, which does not depend upon the explicit evaluation of any partition function. The present method has an element of simplicity and generality that is quite appealing, but this elegance is offset by a certain degree of abstraction that may (or may not) make it less convincing. It is mostly a matter of taste and experience as to which method is preferred.

We begin by noting that the partition function $Z(\beta, V) = \Sigma\, e^{-\beta E_k(V)}$ is a function of both β and the volume V (for simplicity the other mechanical variables, such as electric fields, will be assumed to be constant). The dependency on the volume is due to the fact that the energies of the microstates E_k are functions of V (see, for example, problem 11 in Chapter 2). Now consider the infinitesimal change in the ln Z when β and V are changed by an infinitesimal amount

$$d(\ln Z) = \left(\frac{\partial \ln Z}{\partial \beta}\right) d\beta + \left(\frac{\partial \ln Z}{\partial V}\right) dV$$

Since $U = -(\partial \ln Z/\partial\beta)$, we can write this as

$$d(\ln Z) = -U\, d\beta + Z^{-1}\left(\frac{\partial Z}{\partial V}\right) dV \qquad \text{(C.1)}$$

Now

$$\frac{\partial Z}{\partial V} = \frac{\partial}{\partial V}\sum e^{-\beta E_k(V)} = \sum -\beta \frac{dE_k}{dV} e^{-\beta E_k}$$

and since $(dE_k/dV)\, dV = dE_k$, Equation (C.1) becomes

$$d(\ln Z) = -U\, d\beta - \beta Z^{-1} \sum dE_k\, e^{-\beta E_k}$$

or finally

$$dU = \beta^{-1}\, d(\ln Z + \beta U) + Z^{-1} \sum dE_k\, e^{-\beta E_k} \qquad \text{(C.2)}$$

The present method depends on first identifying the term

$$Z^{-1} \sum dE_k\, e^{-\beta E_k} \equiv \sum dE_k\, P(E_k) = \overline{(dE)} \qquad \text{(C.3)}$$

which is the average value of the change in the energies of the microstates resulting from a change in the volume. This should not be confused with $d(\bar{E}) \equiv dU$, which is the change in the average energy of the system (due to both thermal and mechanical changes). In fact we can write

$$dU = d\left[\sum E_k P(E_k)\right] = \sum E_k\, dP(E_k) + \sum dE_k\, P(E_k) \qquad \text{(C.4)}$$

and we see that (C.3) is just the second series in (C.4). From this point of view the change in the internal energy naturally divides into two parts: one in which the probabilities change but the values of the E_k are unchanged (see Figure C.1); the other in which the E_k's are changed but the probabilities do not change (see Figure C.2) (bear in mind that these are infinitesimal changes). The second contribution to the change in U (Figure C.2) coems about solely because of mechanical changes (e.g., changing the volume). Therefore, it seems reasonable to identify (C.3) with the *reversible work* that is done on the system, $-(dW)_{\text{rev}}$. Exactly *how* reasonable this identification may appear to you probably depends on your experience (and the a pos-

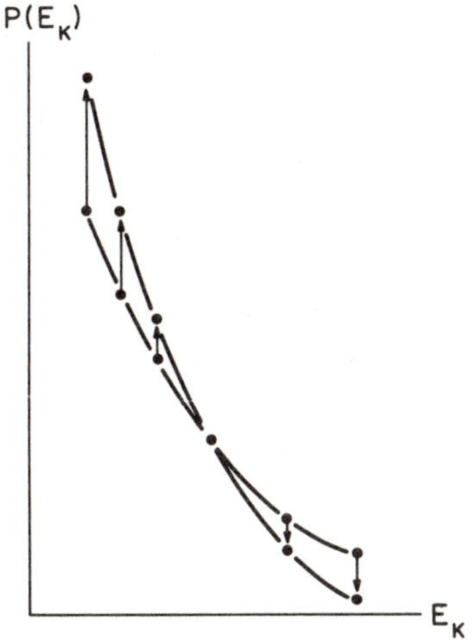

Figure C.1

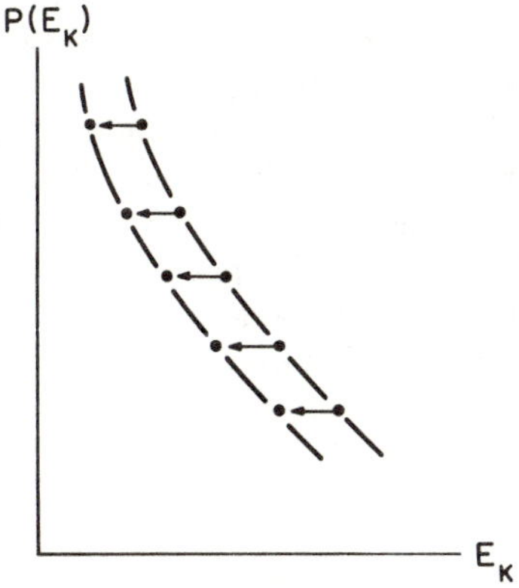

Figure C.2

teriori knowledge that this is, in fact, the right identification!). In any case, if one accepts the fact that (C.3) should equal $-(dW)_{\text{rev}} = -p\,dV$, then the rest is easy — for then (C.2) becomes

$$dU = \beta^{-1}\,d(\ln Z + \beta U) - p\,dV$$

Comparing this with the thermodynamic equation

$$dU = T\,dS - p\,dV$$

leads to the conclusion that β^{-1} is proportional to T, or

$$\beta = \frac{1}{kT}$$

which is the desired result.

D. *CONCERNING THE ENTROPY CONSTANT*

Unfortunately, many books on statistical mechanics pay very little attention to the arbitrary constants that can be included in the identification of the state functions U and S. This omission would indeed be trivial if it were not that, because of these omissions, some writers then obtain nonexisting paradoxes, which are in turn resolved by "correcting" classical statistical mechanics — the implication being that classical statistical mechanics is inconsistent with a very general property of the entropy. Although this erroneous conclusion was pointed out many years ago,† it continues to be perpetuated by even some of the best physicists.‡ It seems worthwhile, therefore, to examine this point briefly.

As noted in Chapter 3, only the *difference* in S (or in U) between two thermodynamic states has any physical significance. In other words, we cannot measure the absolute value of S or U but only the change in their value. The identification of U with the average energy $\bar{E}$ gives rise to no problem, since the energy E is itself undefined up to a constant, so that by changing this constant we can change the value of U as desired (problem 4, Chapter 3). The place where all the difficulty arises is in the identification of the entropy S (and related functions, such as F and G). Consider a closed system (N is a fixed number) for which we have the identification [Equation (32), Chapter 3]

$$dS = k\, d(\ln Z + \beta U) \tag{D.1}$$

or

$$S = k(\ln Z + \beta U) + \text{constant } (N) \tag{D.2}$$

This "constant" may have different values for different values of N — a point which is frequently ignored — but whatever its value, it cannot be measured. However, if this constant is ignored, and S is required to satisfy a "very reasonable" requirement, then it is not difficult to arrive at a "paradox" (which is closely related to the so-called Gibbs paradox).

To illustrate this point, consider a classical perfect gas which has the partition function [Equation (29), Chapter 3]

$$Z = V^N\left(\frac{2\pi}{m\beta}\right)^{3N/2} \tag{D.3}$$

Now if we drop the constant in (D.2) and substitute (D.3), we obtain

$$S = Nk \ln V\left(\frac{2\pi}{m\beta}\right)^{3/2} + k\beta U \tag{D.4}$$

The usual argument then goes something like this:§ "If N and V are both doubled in value, then S should also double in value because it is an extensive thermodynamic function (as is U). But the expression above does not have this property [because of the term $\ln V$ in (D.4)], hence something must be wrong with the classical expression for Z. Therefore, there is something wrong with the way we counted microstates in classical mechanics, etc., etc." The trouble with this argument is that it introduces the require-

†O. P. Ehrenfest and V. Trkal, *Ann. Physik* **65**, 609 (1921).

‡Erwin Schrödinger, *Statistical Thermodynamics*, 2nd ed. (New York: Cambridge University Press, 1952), pp. 59–60.

§See Schrödinger, *op. cit.*

ment $S(2N, 2V) = 2S(N, V)$ on the entropy (which involves a change in the number of particles) and at the same time ignores the "constant." If one sets the constant equal to $-N \ln N + N\times$ (another constant), then the resulting expression for S indeed satisfies the condition $S(2N, 2V) = 2S(N, V)$ (problem 8, Chapter 3). Thus the whole "paradox" has nothing to do with "bad" statistics [or the wrong value of D_N in (13b), Chapter 3] but simply involves an erroneous identification of S in terms of Z. The cry of "foul" may be heard when the constant is set equal to $-Nk \ln N$, on the grounds that this is somehow cheating. In reality the cheating (if any) occurs when one requires that $S(2N, 2V) = 2S(N, V)$, for there are *no physical grounds for such a relationship.* This is simply a *convention.* The only extensive property of S that can be measured is $dS(2N, 2V) = 2\,dS(N, V)$ (where N is fixed) — that is, the *change* in the entropy is an extensive quantity. *This* property *is* in fact satisfied by (D.3) when substituted into (D.1). Therefore, the classical partition function satisfies all of the measurable extensive properties of S, and the additional convention can also be satisfied by an appropriate selection of the constant in (D.2). An alternative method for satisfying this convention is to select D_N in Equation (13b) to be different from unity [problem 8, part (b)]. This is illustrated in problem 6, Section 8, and problem 13, Section 15, of Chapter 4. Although this procedure may be more appealing, it is not a necessary procedure. Put another way, *the extensive nature of S does not imply that particles are indistinguishable.*

All this because of a constant!?

E. *PARTIAL SOLUTIONS OF THE PROBLEMS*

CHAPTER 1

1. (d) None are independent, *assuming* equal probability.

2. (a) Twelve points. *Assume* equal probability. (b) $P(A) = \sum_{i \subset A} P_i = 6 \times (\frac{1}{12})$ (c) A and B are not independent. (f) $P(C \cup A) = \frac{2}{3}$, (g) $P(C|A) = \frac{2}{3}$ (h) $P(A \cup E) = \frac{2}{3}$, (i) A and E are independent.

3. Six points in the sample space: (Ad), (Ap), (Bd), . . . , where p = perfect and d = defective. $P(A|d) = \frac{25}{69}$.

4. (b) 0.005, (d) $4.27 \times 10^4\,dx$ and zero.

5. (a) 1.77

6. (c) $(s_0\,e)^{-1}\,\Delta s$

7. (a) x_0, (b) $(2\lambda^2)^{-1}$, (e) *Not* Monday

8. (a) Not independent, (b) Independent

9. (c) $(c/\pi)^{1/2}\,e^{-cv_x^2}\,dv_x$

10. (a) $t_m \simeq 51$ and 74, (b) $\bar{t} \simeq 66.2$, (c) $\sigma(t)/\bar{t} \simeq 0.18$

11. Var (III) is the largest; H(I) is the largest.

12. The *difference* in their uncertainties is $\ln(\frac{4}{3})$ — *assuming* equal probabilities.

13. (a) $H = \frac{1}{2}[1 + \ln(2\pi\ \mathrm{var}(x))]$, (b) $H = 1 + \frac{1}{2}\ln[\mathrm{var}(x)]$

14. (a) $\lambda = (e\pi)^{1/2}$, (b) $\ln f(x_m) \geq -H$.

CHAPTER 2

1. 3.9412 joules

2. (b) 8×10^{-17} joules

3. Note: $d\mathbf{r} = d\mathbf{r}_+ - d\mathbf{r}_-$

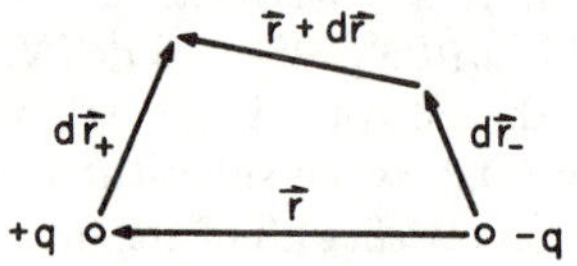

4. $\alpha = q^2/\kappa$

5. (a) 1.24×10^{-25} joules, (b) 1.2×10^{-22} joules

7. (c) $\bar{\phi} = \beta^{-1}$

8. $\bar{\phi} = \beta^{-1} - p\mathcal{E} \coth(\beta p\mathcal{E})$

9. (a) $\kappa = 123.8$ ergs/cm², (b) $\nu = 9.7 \times 10^{11}$ sec^{-1}

10. (b) One diatomic molecule instead of a *pair* of O_2 molecules.

11. $\epsilon_{\mathrm{vib}} > \epsilon_0$; the harmonic approximation?

CHAPTER 3

1. (a) $C = 0.458$, $\beta = 0.424$, (d) $D = C^2$

2. (a) When $T = 200$°K, $P(0) = 0.025$; when $T = 400$°K, $P(10^{-20}j) = 0.0177$ (b) When $T = 400$°K, $P_\ell(0) = 0.324$, (c) In one case it is 37.34.

3. (c) $Z = 10.9$, (d) $E_m = 4.8 \times 10^{-14}$ erg.

4. $U + E^0$; only the difference in the internal energy can be measured.

5. $\mathfrak{z} = \int_V d^3r \int_{-\infty}^{\infty} d^3v\, e^{-\beta\epsilon(\mathbf{r},\,\mathbf{v})}$, proof?

7. (a) Perfect diatomic gas, (b) van der Waals gas (diatomic?).

8. (a) $C(N) = -kN\ln(N) + Na$, where a is some constant. (b) $D_N = (\alpha/N)^N$, where α is some constant.

11. (a) $Z(8V_0) > Z(V_0)$, (b) $P(E_k)$ are equal in the two cases for the same values of k (not for the same values of E_k).

12. Note that $\sum_{\text{states}}(\ldots) = \sum_{\text{levels, } E}\{\sum_{\substack{\text{states} \\ E_k = E}}(\ldots)\}$

13. $\sigma(E) = 2.8 \times 10^{-10}$ joule.

CHAPTER 4

2.1 $N = 3.22 \times 10^6$

2.2 (a) $v_m = (2kT/m)^{1/2}$

2.3 (c) 4.0×10^4 cm/sec for N_2.

2.4 (a) $\eta(\epsilon) = (2N/\sqrt{\pi})(kT)^{-3/2}\sqrt{\epsilon}\, e^{-\epsilon/kT}$, (b) $\frac{1}{2}mv_m^2 \neq \epsilon_m$, (c) 1.167

2.5 $v_{\mathrm{rms}} = (3/2)^{1/2}\, v_m$

2.6 $2\,\bar{v}/e$

2.7 For a speed greater than v_0, the fraction is $1 - \text{erf}(x) + (2x/\sqrt{\pi})\ e^{-x^2}$, where $x = v_0/v_m$.

2.8 0.8427

2.9 (a) 0.19

3.1 $p_1/p_2 = (T_1/T_2)^{1/2}$

3.2 79.2, 19.95, 0.85 per cent.

3.3 $2kT$

3.4 (b) $(v_x)_m = (kT/m)^{1/2}$

3.5 The attractive forces slow down the molecules as they approach the wall. So?

3.6 (a) $\hat{g}\ (u) = 2u\ e^{-u^2}$ if $u \geq 0$, (b) $\hat{h} = 0.5954$, (c) Equilibrium? Spatial uncertainty?

4.2 821

4.3 They compare favorably. Note, however, that the atmosphere is not isothermal. The temperature of 240°K represents the mean temperature over the region $0 \leq z \leq 100$ km, and also the local temperature at 100 km.

4.4 The integral diverges. Is there an equilibrium state?

4.5 1.22×10^{-16} erg/deg.

4.6 $(2N/V\sqrt{\pi})\ C\ e^{+(Cr/L)^2}/\text{erf}(C)$, where $C = mL^2\omega^2/2kT$, and $V = AL$.

5.1 If $100°\text{K} \leq T \leq 600°\text{K}$, then $1 \leq e^{\beta\epsilon_s} \leq 1.23$ for Ne.

5.2 $a = 2 \times 10^5$ cm^6-atmos/mole^2, $b = 20$ cm^3/mole.

5.3 $(r_a/r_s)^3 = 6.44$, $\tilde{B}(250°\text{K}) = -28.7$ (compute other values).

5.4 $a = (16\pi/3)\ (r^*)^3\epsilon^* N_0^2$.

5.5 $$(2\pi/3)(2r^*)^3 \left[1 - \sum_{n=1}^{\infty} \frac{(\beta\epsilon^*)^n}{(2n-1)\ n!}\right]$$

6.1 Quadratic variables?

6.2 (b) $\overline{(z - \bar{z})^2} = kT/\kappa$

6.3 Temperature independent.

6.4 $\dfrac{1}{\beta h\nu} + \dfrac{1}{2} + \dfrac{\beta h\nu}{12}$

6.5 One ratio is 0.58.

6.6 For large n, $\Delta\epsilon_{\text{trans}} \simeq h^2n/4mL^2$ (proof?). For a typical value of n take the most probable velocity, so $nh/2mL \simeq (2kT/m)^{1/2}$ (derive). Then one finds $T \simeq 2.4 \times 10^{-16}$°K (i.e., fantastically small).

6.7 For Cl_2, $\Theta_{\text{vib}} = 810$°K.

7.1 For CO_2, $C_{\text{vib}} = 0.83\ kN$. The abnormal value is related to the rotational motion (how?).

7.2 In the approximation (5.21), $C_V = C_V$ (perfect).

8.1 (b) 12 microstates

8.2 $Z_{\text{dist}} \neq \frac{1}{2}\ Z_{\text{indist}}$

8.3 Degenerate

8.4 Degenerate

8.6 (b) $D_N = C^N/N!$, where C is some constant.

9.1 $\Theta_{\text{elec}} \simeq 2.17 \times 10^4$°K

9.2 $C_V = 3.031\ Nk$

9.3 $(C_V)_{\text{elec}} = Nk\ (y/\cosh y)^2$, where $y = \Theta_{\text{elec}}/2T$; $C_V = 2.578\ Nk$

9.4 The theoretical value at $T = 1600$°K is $C_V/Nk = 3.205$.

10.2 When $T = 0.08\ \Theta_D$, the Debye model yields $C_V/3Nk = 0.0399$ whereas the Einstein model gives 0.0767.

10.4 (a) Consider $C_V(2°K)/C_V(1°K)$, (b) $\Theta_D(T)$ has a minimum value of 135°K.

10.5 (a) $\Theta_D = 28.6°K$.

11.1 As T increases the force increases. Consider the partial cancellation of the electric field between the charges produced by the alignment of the electric dipoles.

11.3 $b \simeq 3.68 \times 10^3$ cm^3-deg.

11.4 Graphically one obtains $d(\chi_e V/N)/d(1/T) \simeq 8.6 \times 10^{-21}$ cm^3-deg.

11.6 As explained in problem 14 of Chapter 1, only the *differences* in uncertainty have significance in the case of continuous variables.

12.1 4.89°K in one case **12.2** 0.023, if $M = -\frac{1}{2}$

12.3 $\mu_m \simeq 2.64 \times 10^{-20}$ erg/gauss

12.5 (a) $S_m = Nk \ln [\sinh [(2J + 1)y]/\sinh (y)] + Nk\, y\{\coth (y) - (2J + 1) \times \coth [(2J + 1)y]\}$, (b) As $y \to 0$, $H \to N \ln (2J + 1)$

12.6 (a) $U_{mag} = -N\mu_B\, B \tanh (y)$, (c) Maxima? Compare in the limit $y \to \infty$.

13.2 (a) $pV = (\frac{1}{3})U$

13.3 (a) $T = 1.41 \times 10^{5\circ}$K for one atmosphere pressure.

13.5 (c) 5800°K **13.6** (a) 0.90, (b) 7.2

13.7 $\pi^{-1}\, e(\nu, 3°K) = 1.43 \times 10^{-17}$ when $\lambda = 7.5$ cm.

13.8 The radiation outside this range is unaltered.

14.1 (b) The internal (average) energy.

14.3 As N increases, E *must* increase.

14.4 (a) Show that $\bar{n}_s/(1 \mp \bar{n}_s) = e^{\beta[\mu - \epsilon(s)]}$

14.5 From $dF = 0$, and $d\bar{n}_q = -d\bar{n}_{q'}$, obtain $\epsilon(q) - kT\, [\ln\, (1 \mp \bar{n}_q) - \ln\, \bar{n}_q - \epsilon(q') + kT\, [\ln\, (1 \mp \bar{n}_{q'}) - \ln \bar{n}_{q'}] = 0$.

15.1 $\epsilon_F = 1.47 \times 10^{-11}$ erg **15.2** $4.21 \times 10^{-3\circ}$K

15.4 Since $\beta\epsilon_F = 63.75$ (when $T = 10^{3\circ}$K), there is little change except in the *immediate* vicinity of $p = p_F$.

15.5 $(\frac{3}{32})\, (3/\pi)^{1/3}\, (\bar{N}/V)^{4/3}\, (h/m)$

15.6 This is one example of where a "reasonable" theory fails rather badly (by a factor of two).

15.7 (a) $\bar{n}(\epsilon) = (3\bar{N}/2\epsilon_F)\, (\epsilon/\epsilon_F)^{1/2}\, (1 + e^{-\beta(\mu-\epsilon)})^{-1}$, (b) $\sigma(\epsilon)/\bar{\epsilon} = 0.437$

15.8 $(5\pi/3)\, (kT/\epsilon_F)^{3/2}/(2\bar{N})^{1/2}$

15.9 (b) In the degenerate limit, $\sigma\, (N)/\bar{N} = (3/2\beta\epsilon_F N)^{1/2}$

15.11 Zero energy?

15.12 (a) $\alpha = 7.46$ ergs/gm-deg^4, (b) $\epsilon_F = 8.28 \times 10^{-12}$ erg, (c) $m_e^*/m_e = 1.365$

15.13 (a) $\mathfrak{z} = (g/h^3)\, \mathfrak{z}_{class}$, (b) $D_N = (g/h^3)^N/N!$

16.2 $\bar{N}/V$ is not required to have a given value for radiation in a cavity.

16.4 (b) $p = 1.342\, (g/h^3)\, (2\pi\, m)^{3/2}\, (kT)^{5/2}$

INDEX

V

W

Z

Physical Constants and Conversion Factors

velocity of light, $c = 3 \times 10^8$ m/sec
electron charge, $e = 4.80 \times 10^{-10}$ esu; 1.60×10^{-19} coul
electron mass, $m_e = 0.911 \times 10^{-27}$ gm
Avogadro's number, $N_0 = 6.02 \times 10^{23}$ (gm-mole)$^{-1}$
Boltzmann's constant, $k = 1.38 \times 10^{-16}$ erg/deg
gas constant, $R = N_0 k = 8.31 \times 10^7$ erg/gm-mole-deg
atomic mass unit, 1.66×10^{-24} gm
Planck's constant, $h = 6.625 \times 10^{-27}$ erg-sec
Bohr magneton, $\mu_B = 0.927 \times 0^{-23}$ amp-m^2; 0.927×10^{-20} erg/gauss
Stefan-Boltzmann constant, $2\pi^5 k^4/15h^3c^2 = 5.67 \times 10^{-5}$ erg/cm^2-sec-deg^4
1 calorie $= 4.184 \times 10^7$ erg $= 4.184$ joules
1 electron volt $= 1.60 \times 10^{-12}$ erg
1 cal/gm-mole $= 7 \times 10^{-17}$ erg/(molecule)
1 atmosphere $= 1.013 \times 10^6$ dynes/cm^2 $= 1.013 \times 10^5$ newtons/m^2 $= 760$ mm Hg

A CATALOG OF SELECTED

DOVER BOOKS

IN SCIENCE AND MATHEMATICS

RELATIVITY, THERMODYNAMICS AND COSMOLOGY, Richard C. Tolman. Landmark study extends thermodynamics to special, general relativity; also applications of relativistic mechanics, thermodynamics to cosmological models. 501pp. 5⅜ x 8½. 65383-8 Pa. $15.95

APPLIED ANALYSIS, Cornelius Lanczos. Classic work on analysis and design of finite processes for approximating solution of analytical problems. Algebraic equations, matrices, harmonic analysis, quadrature methods, much more. 559pp. 5⅜ x 8½. 65656-X Pa. $16.95

INTRODUCTION TO ANALYSIS, Maxwell Rosenlicht. Unusually clear, accessible coverage of set theory, real number system, metric spaces, continuous functions, Riemann integration, multiple integrals, more. Wide range of problems. Undergraduate level. Bibliography. 254pp. 5⅜ x 8½. 65038-3 Pa. $9.95

INTRODUCTION TO QUANTUM MECHANICS With Applications to Chemistry, Linus Pauling & E. Bright Wilson, Jr. Classic undergraduate text by Nobel Prize winner applies quantum mechanics to chemical and physical problems. Numerous tables and figures enhance the text. Chapter bibliographies. Appendices. Index. 468pp. 5⅜ x 8½. 64871-0 Pa. $12.95

ASYMPTOTIC EXPANSIONS OF INTEGRALS, Norman Bleistein & Richard A. Handelsman. Best introduction to important field with applications in a variety of scientific disciplines. New preface. Problems. Diagrams. Tables. Bibliography. Index. 448pp. 5⅜ x 8½. 65082-0 Pa. $13.95

MATHEMATICS APPLIED TO CONTINUUM MECHANICS, Lee A. Segel. Analyzes models of fluid flow and solid deformation. For upper-level math, science and engineering students. 608pp. 5⅜ x 8½. 65369-2 Pa. $14.95

ELEMENTS OF REAL ANALYSIS, David A. Sprecher. Classic text covers fundamental concepts, real number system, point sets, functions of a real variable, Fourier series, much more. Over 500 exercises. 352pp. 5⅜ x 8½. 65385-4 Pa. $11.95

PHYSICAL PRINCIPLES OF THE QUANTUM THEORY, Werner Heisenberg. Nobel Laureate discusses quantum theory, uncertainty, wave mechanics, work of Dirac, Schroedinger, Compton, Wilson, Einstein, etc. 184pp. 5⅜ x 8½. 60113-7 Pa. $8.95

INTRODUCTORY REAL ANALYSIS, A.N. Kolmogorov, S.V. Fomin. Translated by Richard A. Silverman. Self-contained, evenly paced introduction to real and functional analysis. Some 350 problems. 403pp. 5⅜ x 8½. 61226-0 Pa. $11.95

PROBLEMS AND SOLUTIONS IN QUANTUM CHEMISTRY AND PHYSICS, Charles S. Johnson, Jr. and Lee G. Pedersen. Unusually varied problems, detailed solutions in coverage of quantum mechanics, wave mechanics, angular momentum, molecular spectroscopy, scattering theory, more. 280 problems plus 139 supplementary exercises. 430pp. 6½ x 9¼. 65236-X Pa. $14.95

ASYMPTOTIC METHODS IN ANALYSIS, N.G. de Bruijn. An inexpensive, comprehensive guide to asymptotic methods–the pioneering work that teaches by explaining worked examples in detail. Index. 224pp. 5⅜ x 8½. 64221-6 Pa. $7.95

OPTICAL RESONANCE AND TWO-LEVEL ATOMS, L. Allen and J. H. Eberly. Clear, comprehensive introduction to basic principles behind all quantum optical resonance phenomena. 53 illustrations. Preface. Index. 256pp. 5⅜ x 8½.
65533-4 Pa. $10.95

COMPLEX VARIABLES, Francis J. Flanigan. Unusual approach, delaying complex algebra till harmonic functions have been analyzed from real variable viewpoint. Includes problems with answers. 364pp. 5⅜ x 8½. 61388-7 Pa. $10.95

ATOMIC SPECTRA AND ATOMIC STRUCTURE, Gerhard Herzberg. One of best introductions; especially for specialist in other fields. Treatment is physical rather than mathematical. 80 illustrations. 257pp. 5⅜ x 8½. 60115-3 Pa. $7.95

APPLIED COMPLEX VARIABLES, John W. Dettman. Step-by-step coverage of fundamentals of analytic function theory–plus lucid exposition of five important applications: Potential Theory; Ordinary Differential Equations; Fourier Transforms; Laplace Transforms; Asymptotic Expansions. 66 figures. Exercises at chapter ends. 512pp. 5⅜ x 8½. 64670-X Pa. $14.95

ULTRASONIC ABSORPTION: An Introduction to the Theory of Sound Absorption and Dispersion in Gases, Liquids and Solids, A.B. Bhatia. Standard reference in the field provides a clear, systematically organized introductory review of fundamental concepts for advanced graduate students, research workers. Numerous diagrams. Bibliography. 440pp. 5⅜ x 8½. 64917-2 Pa. $11.95

UNBOUNDED LINEAR OPERATORS: Theory and Applications, Seymour Goldberg. Classic presents systematic treatment of the theory of unbounded linear operators in normed linear spaces with applications to differential equations. Bibliography. 199pp. 5⅜ x 8½. 64830-3 Pa. $7.95

LIGHT SCATTERING BY SMALL PARTICLES, H.C. van de Hulst. Comprehensive treatment including full range of useful approximation methods for researchers in chemistry, meteorology and astronomy. 44 illustrations. 470pp. 5⅜ x 8½.
64228-3 Pa. $12.95

CONFORMAL MAPPING ON RIEMANN SURFACES, Harvey Cohn. Lucid, insightful book presents ideal coverage of subject. 334 exercises make book perfect for self-study. 55 figures. 352pp. 5⅝ x 8¼. 64025-6 Pa. $11.95

OPTICKS, Sir Isaac Newton. Newton's own experiments with spectroscopy, colors, lenses, reflection, refraction, etc., in language the layman can follow. Foreword by Albert Einstein. 532pp. 5⅜ x 8½. 60205-2 Pa. $13.95

GENERALIZED INTEGRAL TRANSFORMATIONS, A.H. Zemanian. Graduate-level study of recent generalizations of the Laplace, Mellin, Hankel, K. Weierstrass, convolution and other simple transformations. Bibliography. 320pp. 5⅜ x 8½.
65375-7 Pa. $8.95

SPECIAL FUNCTIONS, N.N. Lebedev. Translated by Richard Silverman. Famous Russian work treating more important special functions, with applications to specific problems of physics and engineering. 38 figures. 308pp. 5⅜ x 8½. 60624-4 Pa. $9.95

THE EXTRATERRESTRIAL LIFE DEBATE, 1750–1900, Michael J. Crowe. First detailed, scholarly study in English of the many ideas that developed between 1750 and 1900 regarding the existence of intelligent extraterrestrial life. Examines ideas of Kant, Herschel, Voltaire, Percival Lowell, many other scientists and thinkers. 16 illustrations. 704pp. 5⅜ x 8½. 40675-X Pa. $19.95

INTEGRAL EQUATIONS, F.G. Tricomi. Authoritative, well-written treatment of extremely useful mathematical tool with wide applications. Volterra Equations, Fredholm Equations, much more. Advanced undergraduate to graduate level. Exercises. Bibliography. 238pp. 5⅜ x 8½. 64828-1 Pa. $8.95

POPULAR LECTURES ON MATHEMATICAL LOGIC, Hao Wang. Noted logician's lucid treatment of historical developments, set theory, model theory, recursion theory and constructivism, proof theory, more. 3 appendixes. Bibliography. 1981 edition. ix + 283pp. 5⅜ x 8½. 67632-3 Pa. $10.95

MODERN NONLINEAR EQUATIONS, Thomas L. Saaty. Emphasizes practical solution of problems; covers seven types of equations. ". . . a welcome contribution to the existing literature...."–*Math Reviews*. 490pp. 5⅜ x 8½. 64232-1 Pa. $13.95

FUNDAMENTALS OF ASTRODYNAMICS, Roger Bate et al. Modern approach developed by U.S. Air Force Academy. Designed as a first course. Problems, exercises. Numerous illustrations. 455pp. 5⅜ x 8½. 60061-0 Pa. $12.95

INTRODUCTION TO LINEAR ALGEBRA AND DIFFERENTIAL EQUATIONS, John W. Dettman. Excellent text covers complex numbers, determinants, orthonormal bases, Laplace transforms, much more. Exercises with solutions. Undergraduate level. 416pp. 5⅜ x 8½. 65191-6 Pa. $11.95

INCOMPRESSIBLE AERODYNAMICS, edited by Bryan Thwaites. Covers theoretical and experimental treatment of the uniform flow of air and viscous fluids past two-dimensional aerofoils and three-dimensional wings; many other topics. 654pp. 5⅜ x 8½. 65465-6 Pa. $16.95

INTRODUCTION TO DIFFERENCE EQUATIONS, Samuel Goldberg. Exceptionally clear exposition of important discipline with applications to sociology, psychology, economics. Many illustrative examples; over 250 problems. 260pp. 5⅜ x 8½. 65084-7 Pa. $10.95

THREE PEARLS OF NUMBER THEORY, A. Y. Khinchin. Three compelling puzzles require proof of a basic law governing the world of numbers. Challenges concern van der Waerden's theorem, the Landau-Schnirelmann hypothesis and Mann's theorem, and a solution to Waring's problem. Solutions included. 64pp. 5⅜ x 8½. 40026-3 Pa. $4.95

LECTURES ON CLASSICAL DIFFERENTIAL GEOMETRY, Second Edition, Dirk J. Struik. Excellent brief introduction covers curves, theory of surfaces, fundamental equations, geometry on a surface, conformal mapping, other topics. Problems. 240pp. 5⅜ x 8½. 65609-8 Pa. $9.95